LETTERS TO MY WIFE

This transcript of the Finley McDiarmid
letters was made by the son of L. A. Reynolds,
1513 High Street, Alameda, California.

Microfilmed for the Bancroft Library,
April 14, 1956

Finley McDiarmid

LETTERS

TO MY

WIFE

Ye Galleon Press
Fairfield, Washington
1997

Library of Congress Cataloging-in-Publication Data

McDiarmid, Finley.
    Letters to my wife/Finley McDiarmid.
      p. cm.
    ISBN 0-87770-472-4 (hardcase) -- ISBN 0-87770-609-3 (paper binding)
    1. McDiarmid, Finley--Correspondence. 2. Pioneers--California--
Correspondence. 3. Gold miners--California--Correspondence. 4. California-
-Gold discoveries. 5. Overland journeys to the Pacific. 6. Gold mines and
mining--California--History--19th century.
F865.M387 1997
979.4'04'092--dc21
                                                                97-28200
                                                                    CIP

ISBN 0-87770-472-4 HC
ISBN 0-87770-609-3 Pb

TABLE OF CONTENTS

EDITORS NOTE

The reader will note that the McDiarmid letters do not follow perfect chronological order through out the text. Since the letters appeared thus in the original work, the decision was made to keep the original order of appearance without deviation or correction.

Susan Paulson
Editorial Department
Ye Galleon Press
1997

Near Centerville, in Appalouse County Iowa Sunday evening the 12th of May 1850.

Having camped a little earlier than usual I have the opportunity of writing to you. This I do because we are much longer upon the road to St. Jo than we expected to be. We by culpable carelessness broke the front axeltree of the Buggy the next day after we left home; between Galena and the ferry upon the Mississippi four miles out of the city. This compelled us to return again to Galena, where we detained until the middle of next day. About 2 o'clock we reached the ferry; the wind was blowing so hard down the river that he, the ferryman could not make our shore; where we were compelled to remain until morning without a morsel of anything for our horses. Next day about noon we crossed. Tuesday the 7th we reached Iowa City. Wednesday the 8th we reached Washington, the county seat of Washington County. Thence to Bryton in Jefferson County. Thence to Ottomwa, the county seat of Wapello County on the East side of the Desmoins river, and 150 mile from Iowa City. From 30 to 40 miles upon the East side of this river, the country will not suffer by comparison with any upon the globe. In passing through Jackson, Jones, Lynn, and Jefferson Counties we found some tremendously bad roads. Sloughs were very difficult to cross. From the travelling of so many to California from all parts during the wet weather that we have not found any road upon the highest hills, but is cut up in dreadful rucks. From this very rough condition of the roads we have been confined to a slow walk all day. We are within five miles of Centerville where I expect to mail this letter. This is a newly settled part of the state, and what little grain has been raised it has been used by the first California emigrants who had been detained by cold weather and the backward condition of the spring at St. Jo and the Bluffs, until the pockets of many of them have been drained of the last cent. I saw in Galena Mr. Rodgers of the Point just returning from the Bluffs. He says that corn sold during the 3 weeks he was there from $2 to $2.50 per bushel. Hay from $1.50 to $2 per 100 lbs. Since we crossed the Mississippi we have found no grain for less than 50¢ per bushel and we have paid 75¢. For the last days drive we have not heard of a bushel to be had at any price. I am very much afraid that from here to St. Jo we shall be forced to drive the team without grain. If so they will fail fast; as yet there is no grass to sustain them. And no hay in the country. The people upon the public roads through this state have fully skinned the people going to California. They have taken from them the last two years more money for their produce than they have had since they came into the state. We passed 2 dead horses yesterday. The weather has been very cold and dry. We have been directed by a set of villians who are interested in ferries to cross the Desmoins at Ottomwa. Consequently we are a great many miles too far north. We are now 180 from St. Jo and the same distance from the Bluffs; where there is from Wiota at least 150 miles difference in the distances. The way we have come we have travelled 350 miles which should have taken us within 100 miles of St. Jo. The weather has been so very cold that I felt much like the ague. Driving gives me no chance of exercising myself. I have suffered greatly from the headache. The other side of Iowa City a man from McHenry Co. Illis. on his way to California died of inflamation of the head the morning we passed. Constantia, I hardly ask you to be watchful of the children that no accident happen about the mill or water. I wish you would write me in Sacramento City California. Direct your letter in large round letters. If any of my fathers people should write answer their letter, my dear, without delay; and write me the information you get. By writing to me soon and often you can keep me informed of your situation; and thus move from my mind a *burden* of *anxiety* about *you* and the *children*. Write me a letter every two months my dear. Do not neglect this. If you can send the boys to school do so. I will write you as often as I can. Good bye. My love to the boys.

F. McDiarmid

St. Jo Missouri, May the 22nd, 1850

My Dear

I thought yesterday that I should be hurried away from this place without having time to write you a single line. We came into this place Monday afternoon. When I went to the post office for a letter from John P. Tramnell I was told by the postmaster that there was none for me. I then went to all the shipping houses and could learn nothing of my freight. I renewed the search the next morning and after the most tiresome examination of the books and enquiry for Grays Whitneys Sublattes, or Tramnells names I accidently found a man who gave me directions to go to a brick house up from the landing and enquire there. I done so and to my surprise found my freight there except one sack of oats which he paid me one dollar and a half for. With Mr. Rice where the goods were stored I found a letter for me which I have marked *B* containing Mr. Rice's receipt for the goods. I went the third time to the P.O. and told the P.M. that I had learned that a letter had been placed in the office for me. He made another search and found the letter or bill which for your information I mark *A*. The direction of the letter is in an open plain handwriting. Any person except General Taylor's P.M. could have read the name the first glance. But they are poor hands to read writing unless writing in very large *Roman* letters upon Flaggs or Canvas. Our horses stand the trip so far very well. The black muck or marle has cracked the heels, and given them scratches badly; otherwise they have done well. Hundreds of wagons for California are within my sight. I saw yesterday many teams for California both horses and oxen cross the ferry that I would never think of driving from here to any place. A large no. of men started with packs upon their backs on foot for California. I saw a no. of footmen start yesterday. Cagle & Notes left the river for the plains 8 or 10 miles out where he could get some grass for the horses. Monday night, the day we came here, we learned that Baldwin camped about fifty miles from here. I am going over the river again to buy if I can a light old wagon upon which we can carry our grain and provisions for 8 or 10 days until we consume the most of the grain when we will place the balance upon the buggy. We are very well. Every chance of sending you a letter I will embrace. Take care of yourself and the children. Do not work out in the *sun* in the garden. Let John do that work. I need hardly tell you Constantia that it is very disagreeable for me to be away from you. You know my *joys* and my *pleasures*. In great haste

Affectionately yours,
F. McDiarmid

I find this getting to California is no *funny* thing. *My Dear* I wish you were with me; I would feel better although my anxiety would be increased. Constantia, you know that I am too much inclined to neglect to take my regular meals at home. It is more so upon the road from the unpleasantness of getting them. Were you with me this would not be the case. I sincerely commit you and the children to the care of God so goodbye.

F. McD

I have not had the time to write to William or my father from this place. I want you, Constantia, to write to them immediately informing them that we have left in good health. As soon as our horses eat up our grain we shall leave our waggon upon the roadside. Our chance, Constantia, to get there is better than thousands that have gone. Do not neglect to write to my father or William.

LETTERS TO MY WIFE

Tuesday the 23rd, St. Jo Mo.

We are still in the *cursed* place. We dare not put all our freight upon our buggy. I have this morning bought a very light strong waggon for $22. Setting tyre will cost about $2 more. We have agreed to take a very gentlemanly appearing person from the state of Iowa to California from this place for $100. He takes about 200 lbs. freight; and the roads are said to be very fine. It is too much labour for 2 of us to take proper care of the horses and get anything to eat in any proper season after we camp at night. As an average I have not slept more than 3 hours in 24 since I have left home. And still I do not feel sleepy during the day. I am so busy and my mind is so occupied that sleep is driven away from me. The Indians are numerous here we have to watch them during the day as well as at night. This tiresome labour made this connection with F. M. Coleman appear to us indispensable. If I were to go to California every month I would never take this route again. You can get nothing that you want without paying four prices for it. I have found sharpers before, but I have never found before a more heartless set of *scoundrels*. This is hastely written in the blacksmith shop. We want to leave this place today; consequently, you cannot expect to hear from us again until we get to Fort Laramie. *My Dear*, stay in the house and take care of our little children; tell them that I send them my best wishes.

20th of May A.D. 1850 Monday morning.

From Platt the country and roads to St. Jo are exceedingly good. St. Jo is about 600 miles above St. Louis, 60 miles above Fort Levensworth and 160 below Council Bluffs. It contains a few good old fashioned buildings made of brick. There are four churches, and among them the Catholic church which is easily distinguished from the rest by its cross conspicuously elevated above its roof. The population which is nearly 2,000 is made up of people of all colours as well as from all countries. Passed another dead horse. It is proper here to say that I have recorded all the dead creatures that I have seen at the camping places and at the road side. But this estimate will fall far short of the actual number. For this reason. The most of the camping places are off of the road from a mile to 3 and 5 the most of which we pass. It is at these stopping places where the animals die. The same explanation or remark will apply to the people as well as animals. So that the no. of graves that I have taken will fall greatly short of the actual no.

21st Tuesday. We have, after some unpleasant delay, made out to get across the muddy Missouri. During the day there was one continued struggle to get upon the boat first. It was a small steam craft owned by the Mormons. When the quality of the road to St. Jo is compared with the road to the Bluffs the latter has a decided preference. Since Fort Laramie, I have learned from people on the Bluff route that there has been little or no sickness. They noticed only three graves while upon the St. Jo route, I counted on the 12th some 22 or 23 graves all of which from the dates of the headboard were buried that day and the day before. This with the difficulty of crossing or ferrying the Desmoins River, the Iowa, the Scunk, and the Platts give the Bluff route many important advantages.

23rd Thursday. A no. of teams returning have crossed the ferry today; and as a matter of course give very unfavourable news about the cholera and small pox. This afternoon we have left the river, past through the bottom timber of the Missouri to the plains, a distance of six miles, where we get rather poor grazing for our horses, and the best spring water that I have tasted since I left home. Very heavy rain during the night with sharp thunder and lightning.

24th Friday. Clear and dry, excessively hot. We have laid still today that our horses may have the benefit of a little grass which they want very much after standing to grain alone since

Monday. While laying by here we have seen many returning who say that the cholera is very mortal among the emigrants 80 or 100 miles from here. They inform us that the pasturage as far as they have been is good.

25th Saturday. We have left our camping ground this morning with a prospect of beautiful weather and fine roads although diverging in various directions upon a ridge or hogs back through a very high uneven country with little timber. Many of the defiles furnish fine spring water. It must be from the dry, clean elevated appearance very healthy. Were it not from this very healthy appearance of the country, I would feel some uneasiness; for we daily meet teams returning some carrying their sick, other who have buried their relatives and companions upon the plain, without coffins, without any ceremony to commit them to their final home. This afternoon Horatio drove upon a sidling place and broke the hind axel tree of the little waggon I bought in St. Jo. The Sac Indians are numerous; and when they come across a few persons going to California they force them to pay them money for travelling over their land. There is a government agency located thirty miles from St. Jo upon the road. This mission or agency is connected with a beautiful farm spreading over a large delightful piece of prairie. Eight or ten huts or houses compose the village. (Notes to be continued).

My Dear, we are in the midst of some 100 teams and men waiting upon the bank of the Platt to get over. I have just got my ticket. Costs $6.00. We are 130 miles from Fort Laramie and 380 miles from Salt Lake making 510 miles from the fort to Salt Lake. *My Dear* I wish you were with me. Take care of the little children. Goodbye *My Dear*.

F. McDiarmid
All well

Thursday evening June the 20th A.D. 1850
Dear wife,

I begin this letter about ten miles from Fort Laramie. I have thirteen pages of notes hastely penned besides the little pocket memorandum book nearly filled. If I live to get through, I will give you the occurences of our journey as they occured. A great part of our journey has been painful monotony. The least calculated to occupy the mind or interest the traveler of any country that I have ever seen. It is truly the most dreary forbidding part of creation. The Indians will be permitted peaceably to occupy this vast country forever. We have great reason, Constantia to thank and praise God for the almost uninterrupted good health we have had since leaving home; while hundreds upon the road both before us and behind us have been suddenly called to their final home. The 12th day of June we passed 23 graves in a distance of 18 or 20 miles; and from the boards at the heads of the graves we learned that they had all died on that day and the day before. Here is a curious fact that 49 out of 50 of the graves contain people from Mo. a few from Ohio some from Penn. and Ill. and 3 from Wis. They are at Fort Kearney and this Post desirous of ascertaining the no. of emigrants that passed which I have got from the Clerk of this Post.

This is taken June 21st at 4 P.M., 1850. Men; 32,760, Women; 493, Children; 591, Naggs; 7,586, Horses; 20,798, Mules; 6,724, Oxen; 21,418, Cows; 3,185. Deaths of persons reported 90. This falls far short of the number.

Fort Laramie is an ancient place, formerly occupied by the fur trading company. It now contains 150 soldiers and 50 men employed by the government. It is situated upon the west side of the Laramie River, a stream that is subject to excessive rises. Sometimes it is not over a man's boots; at others it is from ten to twenty feet. We had to go up one and a half miles from the ford opposite to fort and ferry in the sciff our goods and then take our teams back to the ford and there cross them. I saw a number of waggons upset by the rapid current of the river while I was there; and was told that 4 men had been drown the three preceding days. We are camped about 1 ½ miles from the ford in the sand hills without any grazing for horses, which are fast failing in flesh. We are about one half way through. I wish my dear that you would write to William or Hugh for me stating where we are, the time we have travelling here, etc. I hope you will not delay any time in writing to them as I know that they will be very uneasy until they hear from me. Direct your letter to Nissourie, Brock District, Canada West. I have not time Constantia to write to them myself; and when I say that I have no time to write a letter you will readily conclude that my idle moments are veryfew. Last Monday we settled with Coleman and are now alone. He would whip and abuse our horses so unmercifully that we were glad to get rid of him. We took him about six hundred miles for $34. He is one of those who appears like a gentleman when at heart he is rotten and a full bred *Black Leg*.

The number of our company that we are now travelling with is 29. Horses 36. Some of the men are from Ohio, some from Chicago, some from Racine, and the balance from our place. Jones, the Welch shoemaker, that used to be in the Cape, is one of the company. One of my eyes is so sore from some unknown cause that I am compelled to keep it closed by my handkerchief from the light. Tedious and tiresome as the journey is I could enjoy myself with some satisfaction were it not *for you* my *dear,* and *my* little *children.* When my mind returns to that subject, my manliness leaves me and at once I am near tears. May God in his infinite goodness permit us to enjoy the sweet society of each other again. It is remarkable that people who appear proper at home when upon this route act like Demons. We have in this country the most firey and thundering tempest or storms that I have ever witnessed. Last Monday night we had a heavy storm, with hail, during which I had to hold the little black mare. When the storm passed over, I laid down and was soon taken with the diareah. I had some difficulty to get it stopped. Graves are less for the last few days than before. The forward emigration will escape the sickness upon the road; and have better grazing. We like some others have too much bacon. We are selling it here for 7¢ per pound. 50 pounds is all we should have taken. I think we will place our freight today upon the little buggy and leave the waggon. This will relieve our horses. It is hard drawing even a light waggon upon these sandy roads. Henry and Adam Helmns together with Cagle and Notts are with us and all well. I am sure from appearances that we shall be put to it to procure grazing for our animals upon this route. I send this letter back to the fort by one of the company. I again committ (sic) you and my little children to the care of God. So good bye. F. McDiarmid.

4th Day of June A.D. 1850 A.M.

My Dear,

We are continually meeting people returning. A no. met us last night who had travelled beyond Fort Kearney. They told us, like all the rest, that we have met, that the Cholera and Small Pox were killing them by the hundreds. Between St. Jo and the Big Blue river a distance of thirty five miles we met a large no. who told us that they were dying with the Cholera and Small Pox at that stream in scores. When we came there I could not learn as the Small Pox was there at all; nor could I learn that there had been a single case of the Cholera. Some had died with the bloody flux and Diareah. So far from the Missouri river it is the highest healthiest looking country that I have ever travelled over. In my notes which I have taken daily from the Missouri river I have noted the no. of graves passed every day, horses dead oxen and mules; and am really surprised to see so few graves; most of which contain people from Missouri and Illis. A curious fact. I expect to catch a chance of sending this brief letter upon a half sheet by some returning person to be mailed the other side of the Missouri. Therefore it is somewhat uncertain whether it will reach you. If it should you know that we are all well. We have not overtaken the Wiota people yet.

F. McDiarmid

P.S. We passed Doctor Duncomb yesterday as report says with his *two wives.*

Wednesday the 26th day of June A.D. 1850

My Dear wife,

We are washing, shoeing horses, fixing waggons, etc. etc. etc. today upon the west bank of the La Bonte river a rapid running stream about thirty feet wide varying from 18 inches to 8 feet deep; with some timber upon its banks. We are 64 miles from Fort Laramie and 65 miles from the upper ferry upon the Platt. Here there is a box prepared for letter which are sent by the government trains to the U.S. free of charge. Constantia, I mailed for you one letter at Centerville, one at St. Jo, Mo. and met the mail carriers before we reached Fort Kearny and gave them a letter for you to be put in the P. office at St. Jo for which I gave them a quarter of a dollar. I also put one in the express office at Fort Laramie for you which office is established by a company who sends the letters away every ten days. They charge 50¢ for every letter in advance. The letters put in the government office at the fort sometimes remain some months, and are never sent until some government train is going out. For this reason My Dear I made choice of the express line. I will now begin at Iowa City, Iowa, and give you a hasty as well as brief history of what has occured since we left that city, which was on Tuesday the 7th day of May A.D. 1850 making seven days including the delay at

Galena and the ferry at the Mississippi. 8th Wednesday we reached Washington the County seat of Washington County. In this village which contains nearly 500 persons there are 4 shops or stores and two taverns. We staid at the stage house which was filled with a set of *blagards* and *loafers* who delighted in using the most vulgar, far fetched, and obscene words and phrases that have ever pained my ears, mingled with the most shocking blasphemy. Such as "you are the man that killed Jesus. You are the man that chased Moses." This place is a disgrace to the honoured, sacred name it bears. Thence to Bryton in Jefferson County. Thence to Ottomwa the county seat of Wappaloose County, on the east side of the Desmoins river and 150 miles from Iowa City, 40 miles to Centreville and 200 miles from St. Jo. Mo.

12th Sunday morning, after taking our breakfast upon bread and chalkolate, we started from our camp for Centreville. Sloughs are bad and the high roads very much cut up. No grain here to be had. We have passed through Jackson, Jones and Lynn counties. Are now camping for the night in a fine grove of timber within five miles of Centreville the County seat of Appaloose County. A newly settled part of the state and one entirely without grain. Passed 2 dead horses today. The weather has been very cold and dry since we left home. We have been directed by a set of *villians* who are interested in ferries to cross the Desmoins at Ottomwa, consequently we are a great many miles too far north. From here they call it 200 miles to St. Jo and the same to the Bluffs. When there is from Wiota 150 difference in the distances.

13th Monday morning. Nothing of any importance has occurred or presented itself today. We have overtaken a team from Shellsburg that started on the 25th of Apr.; with Clark, Bentley and Philip Marks. Passed another dead horse upon the road side.

14th Tuesday morning. We have purchased from one Riskendoff 5 bushels of corn for the christian price of $5. He lives 15 miles west of Centreville. Settlers through here few, ignorant, idle, and dirty; they are too far from civilization and commerce; the source of enterprise.

15th Wednesday morning. We have travelled all day and passed but two or three houses. In Missouri we have been in "Hells Kitchen" or St. John; the county seat of Dodge County. In this county seat there is a well and two wretched looking hovels, which they use for houses. We are now in Mercer county. There is good prairie in both these counties and a sufficient quantity of timber. But poor water. They catch the water or rain from there houses by troves that conduct it into wells. These few warm days have started the grass so that our horses can pick a little from the prairies. The timber begins to look a little green. I have slept two nights upon the corn bags. A hard bed.

16th Thursday morning. During this day we have passed over the worst roads I have ever seen. The land is timbered with Oak, Hickory, Black Walnut and Ash. Clay soil. We have passed Princeton, the county seat of Mercer county. It is 60 miles from Centreville. It has 15 or 20 houses with a population of about 100. It is situated upon the east side of the east branch of Grand River; which we ford. We have crossed the middle fork of Grand River by a ferry. The ferrymans children were upon the bank of the stream like the little Indians without any other clothing than dirty shirts. The people through here have an indolent and filthy appearance generally. Harrison county comes next. Bethany is the county seat; and is 30 miles from Princeton. Beautiful weather. Have passed another dead horse.

17th Friday morning. Three men from this state, Mo., came to our camp last night; they are a packing through with horses. Five or six miles from the ferry we crossed the finest running brook of water we have seen since leaving Wisconsin. Five Deer, the first we have seen gave us a whistling good by and were soon out of sight. Roads better. Passed no timber today. Prairies rolling and high. Passed no houses, no farms this A.M. Near sunset we came to another branch of Grand River. It has a delightful appearance and invited us to camp. Banks timbered with Lynn, Walnut, Buckeye, Oak, and Ash. Just above the ford upon the east side of the stream there is a sawmill, and upon the opposite side a *thing* they call here a *Grist mill*. The *outside* is *logs* not *chinked*. The bed stone is placed upon two beams, and an upright shaft the lower end of which receives the water, and the top end of the same shaft bears up the moving stone. Our Shellsburg travelling companions Clark, Bentley, and Marks have a fishing seine which we drew in the stream for fish; but they escaped under the seine among the rocks. A very little rain during the night. The morning is again delightful. We passed through Bethany yesterday afternoon. It has a very forbidding appearance. It is the county seat of Harrison county. It has eight or ten huts or houses. The place is only remarkable for its mean appearance and the immense number of high stumps in the road. No schools!! No churches!!!

18th of May Saturday morning. About 10 o'clock A.M. we met a three horse team returning after travelling 2 or three hundred miles the other side of St. Jo. Using the old saying that they saw the Elephant and took fright. Their history of the probabilities about getting through was very unfavourable. We have passed over a beautiful country today. Prairies, timber and water finely united. This afternoon we have passed through Gentryville, a village containing a population of about 100. In Gentry County. This village possesses a valuable water power with a large supply of water, solid bold banks, and stone bottom. It is a branch of Grand River. This little promising village is twenty miles from Bethany and forty five miles from St. Jo, Mo. Here we were told that an express had lately come from Fort

Laramie to St. Jo. and on his way counted 300 dead horses. We have found in todays travel a beautiful country. We have passed today several teams for the gold regions; they appear to be in a quandary whether to go on or return back. The P.M. of Gentryville told me that including us there had passed that place for California 1552 teams. No grain to be had at any price. We have nearly enough to last to St. Jo.

19th Sunday morning. A delightful day. While we were taking our dinner Miles Bennett a man from Franklin came up to us - he has two horses and no other freight than a little clothing. He intends to get his provisions at St. Jo and pack through. To be continued. I am afraid My Dear you will be troubled to read my writing. My eyes are very sore and I have to write with gogles. From him we learn that Captain Bull of Platville Grant County Wisconsin is about a day behind us. We have learned from Baldwin Gibson Loveless and others by their writing their names upon the trees that they are 7 or 8 days ahead of us. We have travelled this A.M. 25 miles over prairie without seeing a house upon the finest roads that I have ever rode over. The streams and watering places upon the road are universally used by the emigrants for camping. The most of these places furnish a scant supply of wood. Another dead horse by the road side. In passing through Decalb and Andrew counties nothing occured worthy of remark. We are camping upon a fine small river which is called Platt 15 miles from St. Jo. Here we have overtaken a no. of teams from Benton Lafayette County. They left there the 25th of March. The roads and country for the past two days have been fine. We shall feed the last of our grain in the morning. Heavy rain this evening and during the night. There are also a large no. of teams from Davies County Mo. camping around us bound for California and Oregon.

Continued from the 30th of July 1850 mailed at Salt Lake 5th day of June Wednesday morning. Still raining. Late start.

After travelling four or five miles we came to a grave which from the date upon the headboard had recieved its inhabitant the day before. He was from Mo. by the name of John Russell. About two or three miles further we came to a waggon upon the roadside; their mules tethered out, and two graves close to their tent. We learned that a white woman and a colorered one died yesterday with the Cholera and another lady in the tent just dying. In passing this place I suffered a faint blind feeling which I cannot account for nor describe. Today, level prairie, roads heavy. We are camped again upon the bank of Little Blue. Near our camp there are the bodies of two oxen and two mules. Grass good and plenty of water but of an inferior quality. The appearance and soil of this country for 60 or 70 miles up this stream is as fine as man could desire.

6th day of June. Thursday morning clear and pleasant. Have passed today five graves. Three or four of them were filled yesterday with people from Mo. Here we leave Little Blue for good to the left or south of us. Before leaving this stream we supply ourselfs with wood and water for twenty 5 miles. Three antilope passed our waggons and were fired at by two men; but the report from the guns only hastened their speed and they were soon out of sight. Passed today a large no. of ox teams, one horse team, one dead horse and 2 oxen.

7th day of June 1850. Friday morning very pleasant. I am some unwell with a sick head ache. Two miles from Blue we cross a small sluggish stream; and three miles further we cross another, the last place where wood and water can be had for the prairie a distance of twenty five miles. In passing over this prairie we saw five Elk. Roads are drying up. Grass good. Have passed two graves two dead oxen and one horse. Nine men from Shenany County N.Y. with twenty two horses and mules mostly mules, packing, passed us today. They purchased their horses and mules for $70 a piece and their outfit in Independence Mo. We have passed today a large no. of ox teams and have camped in sight of many more that we will probably pass in the morning. We are camped upon a low Muschetoe bottom of the south fork of the Platte River; and within hearing of the *Drums* at Fort Kearney. I have recovered from the head ache of the morning.

8th Saturday morning. Pleasant again after the sun has risen to move off a very unhealthy heavy *fog*. Today about 10 O'clock A.M. we came to Fort Kearney; where there is stationed parts of three companies; in all about 120 men. The fort is located upon the south side of the south fork of Platt river; upon a *low, level, wet* prairie; a considerable portion of which is now covered with water. There are four frame houses, the balance of the *dwellings* outdwellings, stables and other necessary buildings are made of *mud* or *sod*. It is truly a soddy stinking wet forbidding place. The distance from St. Jo. here is called 360 miles. There is here a good well of water. No wood convenient to the fort. We learn from the P.M. that 7280 teams have passed here this season. Letters can be sent from here to any part of the U.S. by paying five cents. We find it very difficult to gather brush willow and stuff to do our cooking. We ought to have carried wood with us from the two streams twenty five or thirty miles back to have lasted us two or three days travel past the fort. Small willows is the only fuel that can be had here. The water is too filthy to be used without settling or boiling. Today we have overtaken Henry Truitt from Galena; and another team from Freeport. One of the persons with the Freeport waggon is a nephew to Jas. Vanmeter. The water in Platt River is unusually high; the flats and islands are all under water; so that the distance from one shore to the other is from 1½ to 4 miles. When the water is confined to the usual channel flood wood can be gathered from the flatts and islands to supply the cooking wants of emigrants. At Fort Kearney there is a blacksmith shop furnished by government for the use of emigrants; so that companies who have blacksmiths with them can do what work they require. Met the U.States mail. Heavy roads. Grazing good. Passed today about 60 teams. 3 graves. 3 dead oxen and two horses.

1850 June the 9th Sunday morning. Looks very much like rain. Thunder and lightning. We have travelled all day upon a level bottom prairie varying from two to three miles wide enclosed upon the right or north by the river; and high elevated sand banks upon the left or south. No timber to be seen. The excessive fatigue of the journey kills the physickal powers of the traveller, and the barren starved monotony of the country freezes up and kills his intellectual faculties. I can say of this country as Bulwer said of the prairies. ''When I look upon it once I have no desire to gaze upon it again''. Today we have met a large no. of teams returning. Have passed more than 100. Eight graves. Two dead horses and three oxen. Grazing good. Slough water plenty. *Willows* the only wood. This A.M. very cold. P.M. very hot. This place is evidently productive of much disease; therefore we think it wise to spend as little time upon it as possible. For this reason and the want of suitable water to use we have travelled today. Have had to wade into the water for wood to cook supper with.

10th Monday morning. Weather although not clear is more pleasant. We have travelled up the river twenty five or 30 miles over a similar prairie or bottom as that which we passed yesterday. Not a shrub or tree to be seen except some few cotton woods standing in the water from which we succeeded in scaling some pieces from one to do our cooking this evening and morning. We should have carried wood from the last stream we crossed 25 or 30 miles the other side of Fort Kearney; a distance of about 100 miles. The bottom has widened out between the water and sand bluggs to a distance of 6 or 8 miles. We have passed 23 graves, one dead person and another just dying in the same camp. 3 dead oxen and 2 horses. Met a no. of teams returning. At noon grass poor. At night we came upon a patch of good clear grass where we pitched our tent. Passed today 40 teams.

11th day of June. Tuesday morning. Very cold and clear. I was up all night upon watch and suffered much from the cold although I was enclosed in my heavy blanket coat. Up to 10 A.M. we passed over a country similar to that of yesterday when we came to a deep ravine or dry creek with a few scatering cottonwoods upon its banks eighty miles from Fort Kearney. A little to the right of the road there is a good spring of water out of which we filled our cans. We gathered up one days wood. Noon. We passed this A.M. 9 graves and within sight of one dining place a company were digging a grave for another. This company is from Ohio; they tell us they have buried 3 of their men who have died with the cholera today. Passed this P.M. 9 graves and two persons who had just died by the road side. Poor grazing, and no timber. As a general thing we have travelled from 3 to 8 miles from the river; and only once and a while a *lone* tree to be seen upon its banks. Sand hills of all sizes, shapes, and heights upon our left. Have passed today 60 ox teams and an immense no. of loose cattle. 4 dead horses and 2 oxen. Roads good. About an hour after dark we camped with little grass for our horses. Plenty of slough water.

12th day of June A.D. 1850. This morning a large Buffaloe passed within sight of our camp; and soon after we started six more crossed the river and the road about a mile before us. Mr. Truitt put after them upon his pony at full speed. He came within 40 yards of them and fired his horse pistol which put them under full lope. I pursued them with the little black mare for 4 or 6 miles over sand hills and hollows. In going down steep hills they would gain upon me. In ascending the hills I would recover the distance they had gained and some more. But with 1½ mile the advantage which they had in the start I became sensible that I could not overtake them without worrying the mare which I would not do. Over this sandy region the grazing is poor. Occasionally we came to points or bends in the river where we could get water for our stock. Passed this day 14 graves; 4 of them from Philadelphia; 2 from Ohio, the rest from Mo. While taking our dinner a man with his outfit in a *wheelbarrow* rolled up big as life. We asked the gallant son of Massachusetts if he wanted any supplies. ''Sugar'' he said was all he wanted. Mr. Truitt gave him half a loaf; for which he returned many thanks placed it in his wheelbarrow, bid us a friendly good bye and rolled on. No wood but willow.

13th day of June. Thursday. Today we have travelled rising of 35 miles over fine roads, based with small stone and sand; as hard adamant, and as smooth as a house floor. Passed fifty teams in our string; about 100 in all; and over 1000 head of loose stock. 20 graves with the bedding and clothing of the persons lying at their graves. Two dead in tent and one young woman dying in the waggon as we passed. Not a stick of timber to be seen on the distance of 35 miles; and no good water. Our horses from their excessive thirst would drink the yellow muddy water of the river. South or upon our left the ground rises gently like the ocean as far as the eye can see, covered with nothing but large bunches of the Prickly Pear and the wild sunflower. We have reached the ford just at dark; and are told that the water is between four and five feet deep. We have seen in one waggon from Iowa bound for the great *Salt Lake Valley* a few fowls, which put me in mind of home. They were driving some sheep; the first I have seen upon the road. An antilope passed before the horses at full speed. I fired at him, but from the loss of the front sight in my rifle I did not touch him. In the midst of Woodards train which consisted of 34 ox teams; four yoked to each waggon; and four mule teams with four mules to a waggon and 250 head of loose stock we passed our wheelbarrow man. Woodard has taken at Independence sixty men and their freight of 100 lbs. each for California for $135 each paid in advance. Rain in the night with sharp lightening, heavy thunder and tremendous winds.

14th day of June. Friday. At this ford we met many coming back. Now for crossing the Platt, which is between a mile and quarter to a mile and a half wide, varying from two to ten feet in depth; changing by its angry current its bottom continually. The water, like that of the Missouri is yellow with sand; so much so that every pail full when settled will yield a quart of sandy sediment. Not a shrub nor any green thing in the shape of timber within sight of the ford. The bustle began by driving into the river the stock and teams. And this was a sight that would greatly amuse a person in any other place except upon the road to California. The languages from all nations where here dealt out with many blows and more curses to the poor animal. The *women* taking an *active, warm part* in the *game*. We took the box from the lumber waggon and placed it upon the buggy box; lashed it fast, put one half of our freight in and quietly moved into the river with the buggy drawn by the bay mare and sorrell colt. The sand was so mirey and the distance so great that the horses grew very tired before reaching the shore. Horatio took charge of this load and I returned for the other. I placed the same team before the other waggon and crossed the stream the second time without any difficulty. We hastily loaded up our freight and with much pleasure left this place. The country from the river is more rolling than upon the other side; and occasionally has a Chaldron or sink containing water. In one of these sinks near the road a large buffalow was killed a short time before we passed. The emigrants took their slices as they passed. In this sickly place we thought it prudent to avoid all fresh meat, and did not disturb him. We travelled about twenty miles from the river when we reached the north Platt where we camped for the night. The north Platt is fenced in by tremendous rough elevated sand hills. We left our stove upon the bank of the south Platt. They are cumbersome *humbugs*. We passed Richard Reed, a brother to Joseph Reed the merchant in Mineral Point, who told us that Baldwin, Cagle, Notts, Helm etc are one days drive before us. We left Truitt and his company upon the other side of the river. He was some unwell, and talks of getting some *chance doctor* to attend him; if he does he will probably add one to the no. of dead upon the road. Have camped at Ash run on Squaw creek. One spring to the right of the road ten feet. Better one 20 feet. 4 graves and 2 dead creatures. I have an opportunity of sending this journal with three other, and a brief letter by Mr. Alexander of Shellsburg to Sacramento city to be mailed there. Alexander has just received a letter from his wife who says that the children in Wis have been generally sick. I feel uneasy about you and the children on this account. My dear write to me every two weeks, direct your letters in a large heavy hand to Coloma Eldorado County California. Greenwood Valley, Nov. the 12th 1850.

Affectionately yours,
F. McDiarmid

15th day of June. Saturday. Weather fine. Settled with Coleman on account of cruel treatment to our horses. Charged him $34 for boarding and carrying him and his freight 570 miles. Nothing but the elevated weatherbeaten sand bluffs of the Platt River to be seen. *Sandy heavy* roads. We have travelled over a valley today that smells like an old ashery. The surface of the ground has the appearance of leached ashes; and the water, which stands in holes, tastes, and resembles that of lye. The ground tastes like ashes wet; or partially leached. The traveller cannot fail to observe the strong smell that arises from this kind of soil. Nothing to be seen in this country that is calculated to produce a lively feeling or a pleasing thought in the mind of the wearied traveller. There is no scenery of any discription to be seen. The country is fit only for Indians. When we pitched our tent upon the bank of the river the air was sultry and still; and the muschetoes were about to devour us; in less than an hour a cloud appeared upon the horizen in N.W. when the wind began to blow and in ten minutes the offenders were *nowhere*. For a time I did not know but it would upset our buggy. The wind lasted until nearly day break, when the morning came in fair and pleasant.

16th Sunday. Poor grazing. For this reason we travel today. The face of the country like that we passed yesterday. No wood; and none but river water. About 2 O'clock P.M. we came up with Cagle, Notts, Adam and Henry Helms. Passed eight graves. Layin by this afternoon. Learned from one of the company with Cagle who was hunting his horses that two men in a camp we passed ½ mile back died this afternoon; and that 2 were dying in a waggon just below us. Cagle, Notts, Helm and ourselves have had very good health. While we were in camp this P.M. there have passed rising of 100 teams.

17th day of June. Monday morning. Very pleasant and fair. We have had a very dusty, tiresome job to pass those teams today that got before while we were camping yesterday P.M. Grazing some better today. Passed in a defile putting up to the right hand side of the road a good spring of water. Our attention has been aroused today by the appearance of an elevated rock which looks at a distance like a magnificent tower. It is about 300 feet high with its base upon the bluff which is 200 feet high, or above the flatt putting out to the river 4 or 5 miles from the tower. Opposite this tower we see another curiousity, which considerably resembles Brocks monument. Freemont in his notes of this route calls this Chimney Rock. They are very interesting curiosities in this monotonous country. Upon the edge of the bluff to our left we have seen a few scatering cedar. Passed 17 graves, and at one of them a few individuals were taking their last painful look upon their friend and departed companion; and after closing up the grave they joined in prayer. A strange exercise for this country!! At the burial of Col. Thorn where I happened to be present the most obscene, vulgar, and blasphemous language that could be uttered from the filthy mouths of any blagguards were spoken by them while committing his remains to the grave. It was so disgusting that I left them. As long as I live, *this place*, with the frightfully grand, teriffic, storms of *thunder, lightning, rain, wind*, and *hail* that we have felt this night will always be fresh in my memory. It was all that we could do while the storm was so violent to hold our horses. If they had got loose we probably would not have got them again. Getting wet in this cold storm I was soon taken with the diareah; so violently that I suffered much during the balance of the night. I resorted freely to Cartrights Cholera preparation and this morning although feeble I am free from it. At considerable distance from the road we saw five buffaloe. Continued upon another sheet. Greenwood Valley. Nov. the 17th 1850. Please send your letters to Coloma, Eldorado County, California.

P.S. I shall mail this at Georgetown where I am going tomorrow. Address your letters in a very large heavy plain hand; as yet I have not received any from you.

I remain affectionately yours etc etc.
F. McDiarmid

18th day of June. Tuesday morning. Pleasant again. No wood today as usual, nor water except the muddy Platt, which we are forced to use. Large elevated monuments, cones, towers, and garrisons of rock are curiously formed from the left bank of this river, by the decomposition and sliding of the large mass of loose material about them, leaving these weather washed, impenetrable, uprights, pillars, and monuments, the only things seen here that can excite wonder in the mind of the gold pilgrim. For the first time since we passed the Missouri we have seen a few Indians and squaws. They are of the Sous tribe and appear friendly. Some can talk english so that we understand them. Their dress is much superior to the Socks and Foxes which we saw at St. Jo. They are a much cleaner and better looking Indian. Roads much better less sandy. Grazing better. By the appearance of another storm about 4 P.M. we were hastily compelled to pitch our tent; and before we were prepared the rain was dashing upon us. Passed 7 graves.

19th day of June. Wednesday morning cloudy. At a distance these bluffs appear like the ruins of some ancient city whose buildings are studded by pillars and columns of the most varied and magnificent kind. Their tops are covered with beautiful rich scenery of Spruce and Cedar. Fine roads today; thus we have travelled 28 miles. 14 of which brought us to where the road raises the bluff and leaves the flatt for 14 miles more; where the road again comes down to the river, where we camp for the night. The bluffs are about 250 feet high. The ascent of the road is long and gentle to the top of the hill where there is a small spring of water and an Indian village containing about 100 Indians of the Sous tribe; they appear friendly to the emigrants. The Socks and Fox Indians own the ground between the Missouri and Blue River. From the Blue to the Platt river belongs to the Pawnees who are now at war with the Sous for a strip of land lying west of Platt river. The Pawnees are hostile to the whites. Grazing poor. Sickness is very severe. Passed today 12 graves and 3 dead persons not yet buried. Truitt has this P.M. caught up to us.

20th day of June. Thursday. Some rain during the night. Fair this morning. Nothing of importance has occurred today. Road in some places good and in others sandy and hard for the teams. We are taking upon the bank of Lye Creek, nearly the size of the East branch of the Piakatonica and takes its name from the quantity of alkalie which the water contains. We have been advised not to use the water even for our horses, until the road approaches or strikes the base of the bluffs near the Indian trading post. Even here the water will if put to the tongue cause it to smart like lye. This trading post contains two buildings, one made of mud, and the other small poles with a shed roof and fifteen or twenty Indian tents made like a cone. We have passed seven graves. Roads today good. Grazing poor.

21st day of June. Friday. We crossed a muddy stream with a very bad bank to ascend, before we came to Laramies branch a large tributarie of Platt, running upon the south side of Fort Laramie. This fort contains 150 soldiers and fifty men in the employ of the government. The situation of this post is much more pleasant than Fort Kearney. The ground occupied by the fort gently ascends from the river, is sandy and clean, so that floors are dispensed with. There is one genteel frame building, and one long eating establishment made of unburned brick. The emigrants office in which their names are taken with the kind and no. of stock is kept in a tent by the hospital. In crossing we were forced to go up the stream from the ford one and a half miles opposite the fort, and get the government sciff and ferry our freight over. Then return to the ford and swim our teams; which we done by placing stone in the boxes of our waggons to keep them touching the bottom, thus prevent the waggon from going round with the current and drowning the horses. We had to go nearly down with the current; bearing gently across the stream. Some of the ox teams of 5 or 6 yoke of cattle when the leaders touched bottom they hawed to when the waggons were upset in an instant. At all these streams many stock are drowned with some men. A large no. or fresh graves appear in the graveyard at the fort. We camped about two miles from town for the night, repacked and let Adam and Henry Helms have our little waggon. Theirs having broken.

22nd of June. Saturday. Travelled eight miles to grass where we camped to remain over Sunday.

23rd day of June. Sunday morning. Warm and pleasant. This day is spent in airing clothes, shoeing horses, repairing waggons and oiling harness. The last I am doing. Five graves we passed between this and the fort. Seven miles this side of the fort we decended a long rocky hill so steep that we had to tie a rope to the axletree and twelve men took hold of the rope to prevent the buggy from smashing the horses over the mountain. Holding back with this rope and locking both hind wheels we decended with out any accident. The country is sandy, hilly, and barren; producing no vegitation; but an abundance of the prickly pear. Of two kinds the flatt and the round. The latter is covered with sharp elastic quills from two to three inches long.

24th day of June. Monday. We are now within view of the mountains. They are high, rough, elevated spurrs towering far above the clouds. This sight produces a lively pleasant feeling. During the forenoon we have passed several fine running creeks with considerable timber upon their banks. Passed today five graves two dead oxen and one mule. In the afternoon we came into a barren region without grass. We drove until ½ hour after ten in the night and then tied our poor horses to the waggons during the balance of the night.

25th day of June. Tuesday. We are off from the river in the hills and mountains; some of which are difficult to ascend and decend. Upon these mountains there are scattering Norway Pine. Passed six graves today; a much less no. this side than the other side of the fort. We have pitched our tents upon the banks of the Laboute River, a stream about thirty feet wide, and between one and a half to two feet deep. Grazing midling good. Roads yesterday and today very hilly.

26th day of June. Wednesday. Today we are washing clothes fixing some waggons, and shoeing some horses. Many lame crippled cattle and worn out horses are left on the road.

27th day of June. Thursday. We have travelled today 20

miles over a sucession of hills and hollows, without the appearance of grass or water. One of these hills is a solid mass of Plaster Parris and as fine a quality as I have ever seen. Passed six graves and ten dead creatures. We have again come to the Platt river after travelling eight miles through the dry burning bluffs, or "black hills", as they are called.

28th day of June. Friday. Last night we camped among the hills on a stream called Alapialla about the size of the Labonte. Grazing very poor. About twelve miles from Alepialla we crossed Box Elder creek, five feet wide, where we took dinner; while here there came up a violent hailstorm which pelled the horses as well as ourselves heartily with hail about the size of hens eggs. We found much difficulty to hold the animals during the storm. Nearly six miles from Box Elder creek we came to Force Boise river thirty feet wide and two feet deep. Passed four graves and three dead creatures. We have pitched our tents for the night upon the bank of a fine stream called Deer creek. There are good springs of water upon the edges of this stream; also considerable timber.

29th day of June. Saturday. Having a midling good grazing spot we concluded to rest our teams today.
30th day of June. Sunday morning. Rain during the night. This A.M. has been so cold that we could not keep ourselves warm while travelling with our over coats on. Fourteen miles from Deer creek we decend nearly a perpendicular bank and cross Mud creek. Here grasshoppers and crickets as large as chickens and as clumsy as snails cover the ground. They must fat on sand, for there is nothing else here. No graves two dead horses. Todays travel has been over the barren bluffs of Platt and we are again camped upon its banks fourteen miles from the upper ferry where intend to cross.

1st day of July. Monday. Pleasant this morning. We reached the ferry at noon, and found there one hundred teams jamed in close by the boats or scows. These boats were connected by short pully ropes to large ones suspended across the river and fastened securely at each end to posts planted in the banks. The boats were of sufficient capacity to receive two small waggons and four or 6 horses at a time. We crossed for $7.00 the price is now $5 for a light waggon $7 for an ox waggon .50¢ a head for horses and $1 a head for cattle. It has been until a few days past double this price. Many people go 3 miles above and there swim their stock over. This is a loss of six miles travel besides being attended with much danger from the swiftness of the stream and the difficulty of getting out of the river upon the banks; which are washed under with the water. If stock miss the place which is cut down in the bank for them to get out they are soon carried down by the current and drowned. Scarcely a day passes but some people in trying to save their stock are drowned. One of the teams in company with us crossed the lower ferry, and in travelling close to the river below the crossing they saw a man lodged upon a drifted tree in the stream. One of the men swam out and towed him in. There was nothing about his person that could give any information where he was from or who he was. They buried him the usual way of burying people upon this road by digging a *shallow small* hole and rolling him in. It was Gibson an Englishman from Shellsburg that was drowned in Big Blue river, mentioned in my former letter. No grass about the ferry. Twelve miles from the river we passed, to our left, an Alkalie lake; around which, there were a large no. of dead cattle and horses. This water is sure to disease whatever drinks it. Two miles beyond this lake, without grass, we have tied our poor hungry horses to the waggon while we have, like wolves, boroughed up in the warm sand for the night.

2nd of July. Tuesday. In travelling from our sand beds twelve miles we have come to the Willow springs. Here, to our right, is as good cold water as can be found on the globe. In this days drive we have passed several alkalie lakes or ponds; all of which are surrounded with dead stock. Four miles west of the Willow Springs we again camp without any grazing for the teams. This P.M. we have had the good luck to get an excellent blacksmith from Iowa to shoe our horses for 12½¢ a shoe. We have passed some curious ranges of elevated rock, rough, rocky, and hilly roads since leaving the river.

3rd day of July. Wednesday morning. Pleasant with some appearance of rain. The road today has been over a heavy sandy desert. We could scarcely see our teams with sand and dust, which in places is two feet deep. We passed an ox and cow team giving out; they were driven by two Dutchman one upon each side pounding the overtaxed, starved and famished animals with clubs until the abused brutes failed to regard the blows. I persuaded them to let their team stay until evening; give them what water they had in their keps, then slowly and patiently move them over the desert eight miles further. Many other teams were nearly as bad. I have counted upon this bad piece of road ten creatures left alive some unable to get up and fourteen dead ones. We have camped four miles west of Independence Rock, upon the north side of Sweet River, and near two large lakes of alkalie. This stream is about the size of Fever River at Galena although not so deep. We found around these lakes grass, although dry and crispy where we secured our horses to prevent them getting to the water during the night. The first grass of any consequence for 6 days.

4th day of July. Thursday. In the morning I observed the horses all appeared dull and sick. While I was occupied in ascertaining this cause of this general sickness, I accidently inhaled their breath which was like smelling a bbl. of saleratus. The ground is covered with this stuff consequently the grass contains a large share of the poison. We soon gathered them up to our waggons where they were kept until we found two or three strips of fine fresh grass with beautiful spring water in the gorges of the mountains six miles from the road. It appeared that the poor creatures could not drink enough water. For the benefit of the horses we shall remain here until monday the 8th. This Independance Rock is so named because it stands upon the bottom ground of Sweet Water River; and has no connection with the mountain or high ground. Is about 300 feet high with a base which occupies two or three acres. It has an interesting, and truly independent appearance. This Rock is used by California and Oregon emigrants as an annual journal or Ledger upon whose base they have inscribed cut and painted thousands or more names. Some of them are in large Roman letters fifty and sixty feet from the base upon perpendicular places of the rock. It is 176 miles from Fort Laramie 18 miles east of the junction of the roads to California and Oregon. 90 miles east of the south pass. 218 miles east of Bear River. 333 miles east of Salt Lake City. 344 miles east of Fort Hall. The healthy, delightful natural bathing fountain near our camp which we have enjoyed during our stay here, deserves some praise. About 200 feet from the base of the mountain and 200 from the top there dashes out from the rocks upon a smooth horizontal table stone a beautiful stream of spring water. It passes over this piece of natural furniture for the distance of sixty feet, the sun having a favourable chance to warm the water, when it takes a perpendicular fall upon a similar rock which discharges it upon a third one from whose natural water worn spout 15 or 16 feet about us we received this valuable and I may say well relished luxury. The 5th, 6th, and 7th were spent in repairing waggons, mending harness, etc, etc.

8th of July. Sunday morning. In good spirits we take our final leave of this pleasant camping place. Our horses have fully recovered and once more move us over the sand hills at a pleasant gate. The country is sandy and barren, producing nothing but, artemesia, or wild sage, prickly pears and mountain brush about six or eight inches high. While we were taking our dinner there drove up at full speed twelve Indians of the Crow Tribe. They saluted us by saying ''how do do'' and by shaking hands with us all. They were mounted upon fine horses which had saddle marks upon their backs, evidently stolen from the emigrants by these treacherous thieves. They were very lively nimble good looking Indians. Some had shot guns, others bows and arrows. The bows were very stiff and springy. The arrows had upon the ends, coming upon the bow string, wing feathers from birds about five inches long; these feathers assist in directing and steadying the arrow while passing through the air. These feathers are nicely fastened at each end upon the arrow with fine sinnew. The other end had a little slit or saw cut into it, in this, there was a thin piece of steel about 2½ inches long and one half inch broad at the base running to a sharp point which was also securely held in by fine sinew. One of the Indians wore a jacket and pantaloons which was all of his dress; the others wore blankets and brichclouts. They eyed our horses closely. This day has been very cold, and the wind has blown a tempest rolling up clouds of dust, sand, and even small stones that have made our faces very sore. Passed today 4 graves. Five dead horses, and 12 cattle.

9th of July. Tuesday morning. Captain Richert, myself, and two others watched the horses last night anticipating the Indians who did not disturb us however. We have travelled up the Sweet Water River today, over a desert region, as usual, some 20 miles. Sandy, heavy roads. Passed 16 dead oxen and cows. As the cows are used with the oxen in the teams, I shall for the future name them, *creatures,* for the sake of brevity. Five horses and three mules. Two graves. We have passed an unfortunate family today; a woman with six small children, she was smashing through the brush and prickly pears barefooted driving an ox team. Her husband died upon Little Blue. Grazing poor.

10th of July. Wednesday morning. Heavy, sandy,

travelling, over a desert; as usual. I have counted today 23 dead creatures, 6 horses and one mule. An immense number of guns are along here thrown away. The people brake the stocks and bend the barrells to prevent the Indians from using them. Hats, coats, vests, shirts, wrappers, pantalloons, boots, shoes, and socks in great no. scattered upon the road sides; together with Ladies' dresses of all kinds, and what must be very painful to part with, is the fashionable *Bussells* that are here left to rot in the general waste. Perhaps a more fortunate destiny awaites the pretty things. Some proud gallant Indians may find them and ornament their necks with them until told by the emigrants that they are worn by the young fashionable women; when they will tranfer them from the neck to the hips of the squaw. There, as the only dress for the squaw, their history will be perpetuated. These Indians although wild and uncultivated have a great desire to dress like the whites.

11th day of July. Thursday. Last night we camped with a large Morman train who sold the upper ferry upon Platte River, and are now returning to Salt Lake. Captain Little who has charge of this train informs us, that for 90 miles we shall have little or no grass. We are now taking our dinner upon the sandy banks of Sweet Water; which we have forded three times this A.M. Teams are failing fast. Our horses being young, of a close tough breed, and never accustomed to grain, are lively and by far the best condition of any I see upon the road. Cattle fail for want of grass and with sore feet. Many of the dead creatures are skinned, and pieces of the hide tied about the feet of those that are getting sore. The weather is here very peculiar, and I think deserves some notice, as I have never seen anything said upon this subject by any person going to California. We have felt daily, strong west winds since we left Little Blue River. These winds, although gentle there, increased in strength as we proceeded west. The gentle pleasant breeze in the morning would by 10 O'clock and during the P.M. assume the strength and violence of the tempest; and this continue until the middle of the night. This cool messinger from the West meets us earlier every day as we travel nearer to the mountains. So that now by 8 or 9 O'clock A.M. we have to face a powerful wind which sweeps the road of all soft earth and dust; leaving nothing but very sharp, heavy sand and smashed stone, upon a base as hard as adamant. The waggons, in rolling over this road, make a noise similar to that made by carriages upon ice covered with a light quantity of frosted snow. It is this cold wind, coming from the snow clad mountains, which we have seen since leaving Fort Laramie with large bodies of ice in the gorges, that supports the wearied famished animals under the burning heat of the sun, and the many privations which they suffer upon the route. I have shed tears while passing those starved and worn out creatures that have rendered for the benefit of man their last staggering effort; and when unable to do him more service they are left in a hopeless burning desert without food or water to pine away and die!! We find Mr. Newman and many others avail themselves of the advantages of this Morman train. They are better aquainted with the locality of the country, and are better able to find grass; if there be any.

12th of July. Friday. We are passing over a sucession of stony hills and hollows. Captain Little tells us that today and tomorrow will carry us over the worst ground that we will find between this and Salt Lake. For in these two days we will pass over the summit of the Rocky Mountains, where snow upon the hills and north sides of the gorges is to be found in abundance. Passed 3 graves, 16 dead creatures, 4 horses and about as many more left to die by the road side. We have driven up one of these gorges where we find some fresh grass watered by the snow and in from the mountain. The soil here is barren and unproductive. And even with a good soil, rain is necessary to assist vegitation, which is not known here during the summer. Passed one branch of Sweet Water and Quakenasp creek which have plenty of midling good water.

14th day of July. Sunday morning. Last night about an hour after dark our horses took fright and twelve out of 21 either broke their larryetts or pulled up their pinns and away they went over the rocks and stones, our four were part of the twelve. I with Doctor Eaton were the watch during the first of the night with Cagle and several others stood by the fire and among the horses when they started. I sprung for the bay mare's rope about twelve feet from me; but she was too quick, pulled her pinn and put off. Another horse not so fast I caught, and without stopping his speed mounted him and put after the others; but they were soon out of hearing. Some of us were out all night. In the morning I found our four and one of Cagles. Just at night 9 or ten miles off in the mountains we found the others. I have seen today not less than 100 antilope and while I was about to shoot one, a man with a gun upon his shoulder run out from the rocks, evidently much frightened at seeing me, and put off as hard as he could run. This shieness or disposition to escape struck me that he was a horse theif, who inhabited these mountains, and probably had some of our lost horses. I was determined to overhall him and ascertain. I shoved the little bay mare down the mountain, over the rocks at full speed risking a thousand chances of dashing her to pieces and breaking my neck. He would look over his shoulder and steer for the roughest places running with all his might. I hallowed to him, but he heeded me not. I hollowed again, and again, and told him if he did not stop I would shoot him. I at last levelled my gun to shoot when he whirled and was in the act of raising his. I told him not to raise his gun or I would shoot him through. He dropped his arms and fell down with fear and fatigue. I soon found that he had camped six miles from us and was out hunting and had tried to get a shot at the antelope that I was after when I suddenly came upon him. In his surprise he mistook me for an indian; and when he would look back he saw the wolf robe I had upon the saddle, this confirmed his fears, that I was an indian, and intended to kill him. He said he never heard me hollow to him. Thus we were both disappointed. I was soon satisfied that he was no *horse thief,* but an honest man, and he was soon satisfied that *I* was not an *indian.*

15th day of July. Monday. The horses that we have used for hunting the others require rest and we have concluded to try it here another day. Several of the company have gone out a hunting. While I am busy fixing harness, halter, larrietts, and pins. While I am writing Messrs sergent, Fred Underhill, Vanmeter, Townsend of Shellsburg, and Wash Parkison drove up and occupy a share of the grazing ground here. They tell us that their horses at the forks of the Platte took fright, and run into the Buffaloe country. Eighteen they lost; 10 of Townsend's and Vanmeter's, two from Fred Underhill, the rest from men I did not know. We learn that several of the large cattle trains behind us are giving out. Thompson lost from his in one day thirteen head, and the poor men, 60 in no., who have paid him for taking them to California are forced to shoulder their packs and foot it. Grass is so poor that it is uncertain about those large trains getting through. Our hunters have brought in one antilope. p e .

16th of July. Tuesday. We have passed today what is called the Pacific Springs; about which I counted 17 dead creatures, 5 horses and several oxen and cows not dead; but they could not get up. Passed today 30 dead creatures and as many more might be called dead. Roads some sandy, although not bad. Very little grass and poisonous water. 3 graves.

17th of July. Wednesday. We are safely over the summit of the rocky mountains. Nothing to be seen here but the rough elevated mountains, covered with deep patches of snow, and a great supply of prairie sage. We have taken the left hand road by the way of Salt Lake. Fifteen miles from the junction of the roads here upon the north road is the junction of the road to Oregon and California. The Oregon to the right or north. Sublattes cut off to the left or south. The Salt Lake road still south of these. Wash Parkison, Townsend, Vanmeter with their company have taken Sublattes route. Cagle Notts Sergant and Underhill are with us. Passed two streams that are called little and big sandies; one dry, the other, is seven rods wide and two feet deep. Twenty three dead creatures and six horses. We are camped for the night upon the bank of Big Sandy. Very poor grazing.

18th of July. Thursday. We have travelled today 27 miles over the heavy, rough, cobble stone roads. 17 of it without water. The small quantity of grass our horses get through here, is so largely charged with alkalie, which is taken up in the growth of the grass from the ground, that I believe they are nearly injured as much as benefited by it. Stock is failing fast. The face of the country is smoother than which we passed yesterday; and produces nothing but this useless prairie sage, except for wood, which is the only natural material with dry buffalo chips that can be had upon a large distance of this route. We have reached Green River and have struck our tents upon its banks. It is a strong running stream with two ferries. The price ferrying waggons is $5 each. They swim the stock. The grazing is dry and poor; and the ground covered with saleratus.

19th day of July. Friday morning. Mr. Brass, one of our company is sick with the mountain fever, which causes great pain in the head and limbs and frequently turns of derangement. His sickness will detain us in this disagreeable starved muschatoe hole today. We have to grease our horses to keep the Buffalo Natts and the bloodsuckers from devouring them.

20th of July. Saturday. We have crossed Green River after having much difficulty to swim our horses. I was compelled to swim the bay mare on horseback and lead the little black one, when the rest followed in. We are now seven miles from the ferry, taking a hasty dinner upon the last bend of the river that approaches the road. Very little short dirty grass here for our horses. This P.M. we have travelled fifteen miles, without water to Blacks Fork, six rods wide and three feet deep, over rough, cobble stone, sandy roads. Passed 4 graves, 16 dead creatures and 4 horses. Four or five miles from this stream we come to Hams Fork, a swift running stream three rods wide and two feet deep.

21st of July. Sunday morning. Heavy rain during the night and evening. Soon after starting this morning there came in sight about 300 of the Snake Indians who traveled with us this A.M. they had 4 or 5 hundred horses. Some of them loaded with packs, and others drawing, by their sides, the indian's tent poles. They appeared to be moving. Were variously dressed; some with britchclouts and buffalo robes loosely held about them; others were dressed with leggings of antilope, elk, and buffalo skins. Doctor Eaton of our company sold them a rifle and two cups of brown sugar for a small pony. Cagle sold them a Mackenaw blanket for $10. Horatio was offered a pony for his rifle which he refused. They are very sharp traders not easily cheated. Again we come to Hams Fork. And soon afterwards to Blacks Fork again. The main road crosses this stream twice, but from the late rain we could not ford it; and travelled up the stream upon the north side. Are now camped upon its banks. Road upon this side of the stream very good. Plenty of snow to be seen upon the mountains. This stream is hemmed in with elevated sand and stone bluffs of shapes and sizes. Grazing poor. Passed 19 head of dead cattle, four horses and one mule.

22nd day of July. Monday. Some rain during the night and cloudy this morning. From our camping ground to Fort Bridger we have found beautiful roads with the exception of a few steep hills. Five or six miles from the Fort we were overtaken by as heavy a rain as I have ever known. It would have done great credit to any country. As soon as the storm abated we drove on to the Fort. Although this place has the name of a *Fort*, there is no military post here, nor has there ever been one. A few trading houses are all. This place has its name from Col. Bridger who has been in this country for twenty eight years. He is a shrewd *yankee*, has an Indian squaw of the Snake tribe for his *wife*. He must be very fond of sleeping with *Snakes*. Bridger and another man from St. Louis own this establishment, which in appearance resembles a yankee pound. A square is formed with the houses, which are made of hewed logs, leaving a small yard about four rods square in the centre. The doors and holes for windows are all facing this yard. The roofs are covered with a large quantity of earth. North of this block there is a yard about the same size as the one made by the houses. The houses make the south side of this, the three other sides are made of small logs planted in the ground as close as they can stand ten feet high from the earth. These are sawed off, and a hewed piece four inches thick pinned upon the top end of these posts. Bridger is located in a sort of basin or harbor surrounded with mountains and bluffs, and those upon the south side give this place the pleasant appearance of snow every day of the year. Passed three graves, eighteen dead creatures, four horses, and two mules.

23rd day of July. Tuesday. Remained here all day for one of the teams that crossed Blacks Fork, and went up on the south side with much difficulty and nearly drowning two of his horses in crossing the streams that come into this Fork from the mountains. They have got here this evening. Plenty of indians here.

24th of July. Wednesday. Rained some this morning. Col. Bridger says he has never seen such wet weather during the twenty eight years he has been here before. Grass good. From the Fork to Cold Springs 6¼ miles we have passed over the highest ground today there is upon the rocky mountains. Travelling up and down some very stoney, steep, long hills. At Cold Springs there is some timber, but no grass. One and a half miles further we come to a small creek and springs. Poor place for camping. Five and a half miles further and we come to Muddy Fork twelve feet wide. Eleven miles more travel brings us to Coperas or Soda Spring to the left of the road at the foot of a hill. The road here begins to ascend another high ridge which is called the summit; and is 8230 feet above the level of the sea. We have been up in the clouds and rain all day. Roads over these hills slippery. Another Soda Spring.

25th of July. Thursday. Noon we are taking our dinner 1½ miles west of Bear River. It is by far the swiftest stream that we have crossed; and from the immense number of blocks used to raise the waggon beds that lie in heaps upon this side of the river it must have been very high. We put blocks in the centre of the buggy springs, but the water came ½ ways up the box notwithstanding. The box was nearly tight and for the short time we were in, the water had no chance to do us any damage. Some timber and willows upon the banks of this river. Grazing short. Two miles from the river the road passed up a valley varying from one half mile to two wide, in which there is an abundance of fresh grass. The soil of this valley is good, being washed from the mountains and irrigated by the rain gathered upon them. Heavy rain this P.M. Passed today three graves, 18 dead creatures, 5 horses, and two mules. We are camping for the night upon the west side of Yellow Creek. Good grass and fine water in the bottom crossing the road north and south. Roads heavy and slippery from the rain. A.M. very hilly and rough.

26th day of July. Friday. A cold November night. Pleasant this morning. The travelling this A.M. has been one rapid decent from the summit of one mountain to that of another. Several fine springs of water. We are taking our lunch upon the bank of

Echo creek, a mirey muddy stream, hemmed in by immensely high close bluffs, and the road follows the creek for sixteen miles and crosses it nearly as often. In the bottoms of small ravines putting into this stream the coarse grass is very good. More rain this P.M. Today we have found by far the worst travelling we have had upon the mountains. We have been compelled to tie a rope to the hind axeltree of each of our waggons in crossing this stream for 8 or 10 men to hold the hind part of the waggon from pitching upon the team and smashing them out of sight into the mud. Passed one grave. 26 dead creatures, seven horses, and five mules. We have driven twenty five or 30 miles today and are still in this gorge with the huge massy jaws of the mountain nearly closed yawning four or five hundred feet above our heads. In this unpleasant, although magnificent grand place, our horses are turned out among the rocks without anything to eat for the night.

27th day of July. Saturday morning. Started as soon as we could see to harness our horses. And about 9 O'clock we reached Webber River where we took our, well relished, breakfast. This is a fine rapid stream, six or eight rods wide, and four feet deep, with considerable timber upon its banks. Some of our company have caught in this stream a few trout. To the right of us in one of the gulches cutting the main bluff of the river stands a beautiful red monument or pillar of stone, apparently round, and about 150 feet high. Four miles from this we cross the river, and have travelled about twenty miles over very bad roads. Hills, pitches, perpendicular drops, side hill slides so steep that we were forced to tie ropes to the upper sides of the waggons and eight or ten men hold on to these ropes to prevent the waggons from sliding to the bottom of these gulches. Our travel today has been down a creek which we have crossed twenty or thirty times; it intersects another creek called Kanyan Creek; one rod wide, and one foot deep. We are camping for the night at the junction of these streams. 4 dead creatures.

28th day of July, Sunday. Some rain this morning. Today, we have traveled by far the worst roads that I have ever seen a waggon rolled over. Up one kanyan and down another continually. Hills or mountains from one mile to five in length so steep that an animal can scarcely keep a foothold. These kanyans as they call them are narrow gorges, with close elevated mountains; the most of which furnish fine water. In climbing up the last kanyan, that makes from the top of the mountain that overlooks Salt Lake Valley, we have passed over stone smoothly polished with water, from one half foot to three feet in size. Upon the edges of this kanyan which is nearly two miles long there are some beautiful fir and spruce trees. These trees had a *cheerful, pleasant* appearance to *us*. This long, steep, tedious mountain is very dangerous to descend; at the foot of which we find a piece of table ground that has very good grass, where we camp for the night within twelve miles of the *great Mormon city*.

29th day of July, Monday. We leave this stream made by the descending canyon to our left and ascend a smooth steep mountain one mile long to the right, which gives us a nearer view of the valley. Here is fine water. At noon providentially we find ourselves in the *City*. Ignorance and superstition, misery, and want and suffering stalk broadcast in all parts of the valley. With the exception of three or four houses, and the government house is one of them, the rest are 8 by 10 and 10 by 12 made of unburnt brick or clay, and covered with earth. Some use covered wagonboxes for houses. The city is laid out in ten acre lots; three of which make a *ward*, from each ward they elect a *Bishop*. These Bishops compose their *council*, over which their President Brigham Young presides. They have also a judiciary, but its powers I did not learn. The city is located upon the east side of the valley, six or seven miles from the base of the mountains upon a smooth piece of ground gently descending south and west; so that with a small expense they have taken from the mountains in small ditches the water with which they have abundantly supplied every ward in the city with the best of water. Nearly three miles north of the city there comes dashing out at the foot of the mountain a hot sulphur spring. This carried to the city in pump logs and then used for bathing purposes etc, etc, etc. We were told at Platte River by the Salt Lake settlers, that we could, when we reached there feast oursevles upon the vegetable delicacies of the place. But in this our appetites, longing for vegetables, were very much disappointed. I hunted through the city for corn potatoes onions peas and beans. I found none but peas, and they were the size of No. 2 shot. Wheat with the exception of one field which Mr. Sergant said was as good as any he ever saw in Wisconsin poor. Indeed everything poor. The first buildings of this place which are now fast falling into decay, were placed in four direct lines, forming a square for security and military strength. These have small port holes in the outer sides, from which they could shoot. There is no timber in the valley. It is obtained by *men* not *teams* from the mountains with some danger and great labour, because David and Solomon procured their materials for the Temple of Jerusalem from far countries with much expense and labour. they also *herd* their *cattle* here, which, they say, *proves positively* that *they* are *Gods people* as well as Abraham, Isaac, and Jacob who herded their flocks. The people that I saw at work were irrigating their unfruitful garden patches. There is a post office and a printing press here. But I think their *exchanges* are few. Captain Richert and I called upon Captain Stansbury, who has been ordered here by our government to make a survey of this country for information as to the routes from here to California. He has advised us to take the northern route, and risk the grass upon that road rather than the alkalie desert or swamp of seventy five miles on Hastings cut off on the south route. Captain Richert,

Doctor Eaton and many others have sold their wagons and are preparing to pack through upon the southern route. Favourable stories of grass, water, roads, and the short distance upon the south route are passing from mouth to mouth and for the past few days the emigrants are mostly taking that road. The advice of Captain Stansbury and all the Mormons to the contrary notwithstanding.

30th of July, Tuesday. We are still in the valley making enquiry about the roads and listening to the rumours. There is probably not much grass upon one route and many dangers upon the other. Captain Little of this city, with whom we travelled a few days from the Platte River, gave Richert and myself an invitation to dine with him. We, of course, accepted this hospitable invitation, and found his table well loaded with good things which we eat with an eager *skill*, that was clear evidence to him that we relished them well.

31st of July. Wednesday. We are still in the valley. Flour is selling here to the emigrants at $25.00 per cwt. Butter 50¢ per lb. Milk 10¢ per qt. Eggs 50¢ per doz. Bacon 75¢ per lb., none to be had. Beef 12½ ¢ per lb. No groceries in the city. We paid twenty-five cents for a small pole to make a fire, to do cooking by. We are truly in the *fog* as to the route to take from here. Cagle says he has seen several persons who took Sublattes cutt off returning to this place from that route for want of grass. They say that it is impossible to get stock through upon Sublattes route, or the norther route from the city. They have lost ½ of their teams and now intend to try the south road from here, called Hastings cutt off.

1st day of August. Thursday. We have tented during our stay here above the bridge, upon the east side of *Gordon*; a stream about eight rods wide, and from ten to twenty feet deep. The direction of this river is from the south east to the north west; and empties into Salt Lake. Opposite to us a smaller stream empties into *Gordon* which makes nearly an island upon which we kept our horses. During the night they swam over the stream, and have made their escape. It is now some time after dark and I have been hunting them since daylight in the mountains without anything to eat. Captain Richert found them about 2 O'clock P.M. hitched up to a pair of bars in the city. He soon learned that two of them had been at a wheat stack; for which the owner claimed $5. damages. They have a law here that all loose creatures in the city can be taken up and put in the stray pen by these swindlers who get one quarter of the spoils. Their fences are made of long crooked poles, three or four to the panel; separated by pins at the ends bored into upright pole posts. I have seen but *one four legged hog* or *swine* in the city. The great fountain of good fresh water and the sulphur spring are all that I can admire here. The *Mormons* all aspire to be *Solomon*, and have, like unto him, many wives and concubines. Their creed allows them as many as they can support. This *support* is a matter of *comparison*. And I suppose if their wives and children are partially *clothed, uneducated,* and go *barefooted*, they are supported genteely. Brigham Young, now governor, has *30 or more;* and Mr. Sargent says that Carrington has 3, or more, etc, etc, etc.

2nd day of August. Friday. We have had the satisfaction to pass over *Gordon*, on the bridge, by paying $1.25 toll, and are taking Hastings cut off, the lefthand road from the city contrary to the advice of the Mormons and Captain Stansbury. Thus we intend to risk the alkalie swamp or desert. Sixteen miles from the city, in sight of Salt Lake, which is 25 miles west from the town, we find a beautiful spring of very salt water. Two miles further we take dinner there is another, although not quite so salty. Roads fine, and grass dry but plenty. No wood. Large numbers of dead creatures about the city. Salt Lake and springs are now upon our right hand, and the mountain from one and a half to three miles high clothed with some snow upon our left. This lake is a beautiful body of water so strong that a crustation of salt is formed upon the bodies of those who bathe in it; and is so transparent that a small bright substance can be distinctly seen at the depth of twenty five or thirty feet in the water. Near the south part of the lake, 26 miles from the city we are camping for the night, and close to a sulphur spring upon the left hand side of the road upon a small mound. The water, I think, is healthy and not disagreeable to the taste. About ½ miles before reaching this mound spring we passed a small salt water spring to our right. A curiosity that two sorts of water are discharged so near each other.

3rd day of August. Saturday. Five or six miles travel this morning brings us to a small break or dropping in the ground from which a large body of salt water is dashed out. Ten or twelve miles from this place we reach Willow Creek. So called from the quantity of willows on the edge of the stream. This is the first pleasant pure fresh water since leaving the city. Roads dreadful dusty. Abundace of grass here and of an excellent quality. Plenty of dry willows for wood; this is quite a luxury. Passed four dead oxen. We camp upon the bank of this creek.

4th day of August, Sunday. A Montana company share with us in the grass. Mr. Misner, Dr. Griffin's former partner in Georgetown is with them. Jones of Milford is with them. One of the company died of the cholera last night, another is not expected to live this morning. Seven or eight miles travel north of west we came to a number of large salt springs remarkably clear, from these springs we bear to the south up a valley, the mouth of which borders upon the south part of the lake. The road which runs in some places close to the base of the mountain on our left is very stoney, and when the road is further from the bluffs upon the bottom the dust is intolerable from one half to two feet deep. From this bend or point in the mountain we have travelled in a south direction up this valley 37 miles, making the days journey 45 miles without any grass or fresh water. We cut *hay* at Willow Creek, and there filled our cans with good fresh water so that we and our horses have not suffered. We are camping where the road turns west and crosses the valley which is 15 miles. Passed two dead horses and four oxen. Plenty of good grass here and an abundance of fine water.

5th day of August. Monday. We are busy this morning cutting and curing hay for our horses where we will remain during the day. From this place there is no grass for 75 miles. Here is an abundance, and numerous springs of fresh water boiling up in all directions. These springs are very deep like wells, and dangerous for cattle. If they slide into them, they are gone forever.

6th day of August. Tuesday. Started early, and find at the foot of the mountains after crossing the valley 15 miles a large supply of water. Tastes very much of copperass & sulphur. Whiles here a heavy thunder storm came up which lasted three hours. From this copperass spring the road passes over a succession of rough steep hills, by far the most perpendicular of any we have found upon the whole route, bearing north of west for the distance of 40 miles when we gained the summit where we begin to descend similar hills upon the west side of the mountain until we come to the desert, which is level; and the road from one to two feet deep with sand and dust. Twelve miles from the foot of the mountain, or edge of the desert we come to what is called Stoney Ridge. It is a *poor* hogs back mountain with a tremendous *steep stoney* hill to descend.

7th day of August. Wednesday. Large portions of this desert bears a low rough prickly shrub or weed other parts appear like the base of dried lakes; the earth is a sort of ashy substance filled with lye, and a very little rain makes it impassible. One place which we have crossed this morning was made very bad from the rains yesterday, and while resting our teams and taking a bite of breakfast it has rained here some and from appearance more west. Upon the middle and west side of this desert the surface is covered with bright scales of crystalized salt, which resemble by reflection large bodies of water to persons at a distance. There is also another peculiarly beautiful deception caused by this bright substance reflecting the rays of the sun, which magnify, or represent a horse to be ten times taller, and larger than an elephant; and a man to be from fifty feet to one hundred feet high. Some parts are as smooth as a house floor as far as the eye can see. The sun beats down here with indescribable heat. We have travelled the 60 miles and are told here by some whose teams have given out, that we are yet thirty miles from water and grass. They have left their wagons, freight, and families upon this burning desert, and have driven what stock can travel through to grass and water. When they recover, the people will return for their wagons. No alternative for us but to drive our tired horses thirty miles further without water. We gave them the last fifteen or twenty miles back, to avoid the weight of drawing, expecting from information to get water at the edge of the desert or foot of the mountain. This *false* information has produced universal disappointment and with it much loss of stock; and no mouth can tell the amount of human suffering and misery!! We have seen a number of men and women begging and trying to buy more water. Passed twelve dead creatures and four horses. And many more that might be called dead left upon the road. We have travelled all night and have passed an immensely increased number of giving out and dead stock. About 2 O'clock in the night or morning we met two carts loaded with water for the use of those persons and creatures that are back, and are unable to proceed any further. Those who had money paid from one to five dollars per gallon. Those who had none of the *ready rino* received it 'without money, and without price.' Mr. Misner has just got over the desert. He left the balance of his company upon the side of the 15 mile valley. Several of the company have died since we left with the cholera, and others are very low and not expected to recover.

8th day of August. Thursday. We have come to water just at day break. Our horses very much fatigued, as well as suffering excessively from thirst. We intended to lay by today, and rest our horses. Grass very poor. Two loads of water have been sent back today. As Captain H. Stanbury told me this is a very dangerous desert. A stick can be shoved down by the hands into the ground any depth. An ordinary rain would make the road altogether impassible for footmen, say nothing about stock and wagons. Out of the roads there is no *bottom* to this ashy, sticky, stuff. The *travelling* upon this road makes it solid. In a long dry season a smoothe crust is formed, over which light wagons and stock can pass with ease. But this is the result of a very dry hot season. Many people have been unable to get over without assistance of their friends, who have gone back from here with water and fresh teams to their relief. Some have already perished. I was told that a purse of $500 was given for a *cup* full of *water* which he drank, and asked for another which he hastily drank and asked for the third. The taker of the purse told him that for his own safety he would refuse him the third cup full; although he had paid dearly for it. And with a hearty laugh generously gave the suffered back his purse. This is probably one instance out of many to show how *little money* is worth under such suffering circumstances. Thirty eight miles back upon the desert, yesterday evening, we passed Doctor Boyce amd Misses Hall in their wagons without a drop of water. We had a little in our can which we gave them. The teams had given out and were driven loose by Mr. Hall and others of the company, in hopes of finding feed and water in a few miles; when they had to go thirty eight. This false information as to the the precise distance across the desert to grass and water was mostly given by a Mormon who was building a saw mill 25 to 30 miles this side of Salt Lake. He sold to the emigrants, who are generally too ready to grab at any information or receive any man's story, a chart or a map of the road over the desert, marking the springs, feed distance, etc, etc, etc. He sets the distance at 60 miles, where as it is at least 90 if not 100. I here send you a *fact simile* of his map. As you hold this sheet in your hands, our travel was from his mill to the Salt Springs west, then north 45 miles, then west across the valley 15 miles. Then north up to the summit and down to the foot of the mountain 8 miles. Then west again to the hogs back or rocky ridge 12 miles & so on to the 40 mile bend in the road which we expected would lead out to the mountains; but it gently inclined to our right hand and took us 30 miles beyond where we were told that we would find water and grass. A particular description of this map route you will find between the first and eighth of August.

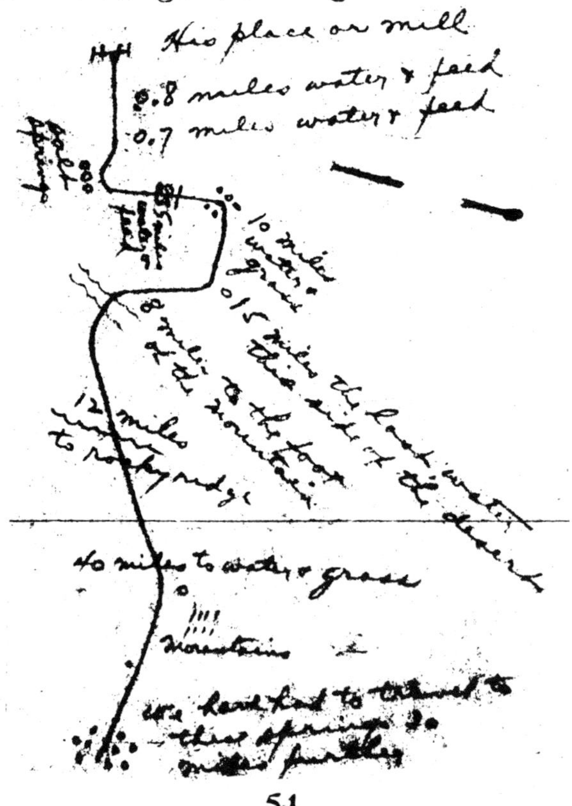

9th day of August. Friday. People coming acrosst this morning continue to give the most alarming intelligence from the desert. Several teams and many pack mules have been loaded with water for those famishing upon the plains. I now learn from Mrs. Hall who has just got through that Mr. Hall reached the wagon worn out and sick. Then she was forced to become an *ox driver*, took the whip and at their side footed it through. Over this last 30 miles the dead horses, cattle, and mules literally cover the ground. And here they are numerous, making a smell that is intolerable. We have moved out to the last fresh water spring this side of another desert of 37 miles to avoid this offensive stink from the dead stock. The grass although cleaner here is very poor.

10th day of August. Saturday. From six to eight miles travel, this morning, between the base of two mountains has brought us upon a plane, over which, we pass when we come to a canyon leading through another range of mountains. At the mouth, or one and a half miles up the canyon there are two small springs; where we had to wait our turn some 3 hours to water our horses. On account of this delay, we cannot pass through the mountains before dark; therefore we think it best to stay all night. I took the horses, buffalo robe, and gun and started up in the mountains two or three miles for grass where I watched them during the night. Notts, Cagle, Sergent & Co. who were in advance of the train went on soon after we came to the springs. They said that they saw some Indians there as they were coming up the canyon. Others said there were two or three hundred of the Shew Shew tribe living but few miles off. However, I did not see any. I counted twelve dead creatures, six horses and five mules at these springs; and several before reaching the spring.

11th day of August. Sunday morning. We filled our can with water and started up the mountain which is about four miles to the top, where we began to descend upon a sloping plane, covered with low sage brush and other useless stuff. This declining plane continued for seventeen miles when we come to water and grass, at noon. Both very good. This is by far the best place for recruiting stock that we have found upon the whole route. The grazing is from ten to fifteen miles long and from 5 to six wide. Laying east of a large range of mountains running north and south. This range we must cross, and to do this the road takes a south direction for a favourable canyon leading into the mountains. As near as I can judge we are now about two hundred and thirty, or forty miles from Salt Lake. *Common talk says* that there is a road or pack trail putting out from this on our left, striking Carsons River, leaving the St. Mary's, or Humbolt entirely to the right. If there is such a road we are determined to take it, as the only hope of getting our horses through. If we find no such road, and are forced to take the old thoroughfare down the St. Mary's River, I am sure our horses must starve. P.M. some rain with thunder and lightning. Passed 7 dead creatures, four horses & two mules.

12th day of August. Monday. We are laying over today for the benefit of our horses, and they require it much. Here we overtaken Sargent, Cagle, Notts, Fred Underhill and George Bigaloe. Cagle, Notts, and Fred have gone into the mountains for game. Have returned with two rabbits. Snow here upon to top of the mountains.

13th day of August. Tuesday. We have travelled down the valley south, or south west, over a very good road for sixteen miles to *warm* springs. Situated nearly in the centre of the valley, about five miles from the mountains upon the west and east. Here we camp for the night. Grazing poor. These springs which continually carry up light particles of earth with the water have made elevated mounds; from the top of which the water is discharged. These mounds are sodded over with grass, and a person can stand or jump upon one and shake the ground for ten or 12 rods. Sixteen feet from one of these warm springs. I tasted the water discharged from another, found it cold and strongly charged with sulphur.

14th day of August. Wednesday. From these springs we turn directly to the right or nearly west; and cross another range of mountains. The distance over variously estimated at from 4 to 8 miles. The road ascending & descending the mountain gradual. Here we come into a magnificent basin or desert surrounded by mountains. The distance from our side of this basin or desert is from 16 to 20 miles. At, or near the foot of another mountain that we cross there is another mound or spring of warm water which tastes strong of sulphur. To the left of this spring south of the road one half mile there is another colder and much better water. At this last spring we camp for the night. Grazing poor. Road over the basin hard and pretty smoothe.

15th day of August. Thursday. Last night *rumour* among the camps told us that the night before there were stolen, from this place, by the Indians, three horses and one ox. But I regarded the report as I had a thousand other stories that prove to be false. However, I went out and watched the horses for two hours after dark, then brought them to the camp and tied them. At 2 or 3 O'clock in the morning I got up, so a number of teams just coming into camp. Fires were burning briskly, people among their stock. The appearance was no ways suspicious of danger; but to the contrary, almost perfect security. I found the watchman upon his duty, who was a Dutchman from Dane County Wisconsin. I then turned our horses loose with those he was watching. A little before day, as usual, I got up and searched through the stock for the horses without finding them, and as I was returning, I was surprised to find a dead horse with an arrow in his side opposite his heart. Another horse belonging to Robert Slaighton of Madison, Wisconsin was carrying an arrow in his shoulder blade. If the arrow had hit him an inch further back it would have killed him instantly. The arrow was pulled out but the flint upon the point was left in. With some difficulty this was extracted. This and the absence of our horses convinced me that the Indians had stolen them. A grey pony belonging to Robert Slaighton was also missing. We hunted the edges of the mountain around the basin for the trail, which we struck about 2 or 3 O'clock P.M. not far from the camp, where they drove them across the road to the north. I followed the trail for three or four miles after Horatio and Slaighton turned back. I am sure from the running of the horses and the right angle turns in the trail, and from the tracks of two Indians that they could not catch the horses, although the little black mare had a loose larryette hitched to her halter and dragging among her legs two rods long, and the big bay one an iron halter chain swinging loose. Without these upon them, I could not catch them. the pony was caught and used in driving ours. I returned for water and something to eat. While here there came up a heavy rain and hail storm. In this unpleasant embarrassed situation we thought it necessary to make some bargain, with some one, now here, to take our buggy or provisions and clothing through. Some asked $200. One offered to take the buggy through *if* he *could* with 300 lbs. for $150, or $50 and the buggy. At last Mr. Hall told us that he would take our provisions and necessary clothing through for the buggy; which offer we accepted. I then tried to get four or five men to go with us in pursuit of the Indians; but received a reasonable satisfactory excuse from them all. Indeed it is painful for me to say that our prospects, bad as they really be, yet they are as fair, and as cheering as the most of those now upon the road for California. They are here with broken down, worn out teams, five or six hundred miles from California and all nearly, and many without provisions. In this alarming situation we failed to raise a sufficient number of persons to pursue and chastise those Indian thieves.

16th day of August. Friday. After throwing away our tent and everything else that we could by any reasonable sacrifice spare, we have started with an old *yoke* of *oxen* before our buggy. This is gloomy, doubtful prospect for California. We raise another mountain of gentle ascent and descent of six or eight miles base when we again come upon a valley of 12 or 14 miles where we come to a fast running stream of water to our right, at the base of another range of mountains. Here we camp for the night. The road over the valley, not bad. Grazing here is very good. Passed 2 dead horses, and one shot through the neck by the Indians still alive. The Artemesia, or wild sage, is the only fuel to be had over these valleys and mountains. The road to ascend the mountains now before us, turns south.

17th day of August. We have passed over another mountain of eight miles, occasionally covered with scrubby cedar. The road not bad. We again come upon a valley of twelve miles; which takes us to another mountain out of which gushes a fine spring of water, where we camp for the night. Grazing fine. Willows & sage brush for wood. Snow here in the gorges of the mountains. Road from the mountain very good. Have travelled today southwest twenty miles.

18th day of August. Sunday. We are laying over today and in the P.M. Captain Richart, and others, that we left in Salt Lake, have overtaken us. Richart was all spirit to return for our horses. He found eight or ten men in camp who were willing to accompany him; although some were entirely out of provisions. In this starved state of affairs I thought it best to bear our loss without making others in any way liable, or subjecting ourselves to any censure in case of accident. After debating the matter for some time, we finally concluded to make every step towards California.

19th day of August. Monday. The road from here crosses

the creek and lays south up the east side of the snowy mountain; from the base of which dash numerous springs of good water which makes our road in some places bad. Plenty of grass along the base of this mountain. Passed ten dead horses. Have travelled about 23 miles, and have camped for the night upon a fine stream with willows upon its bank convenient to the road.

20th day of August. Tuesday. We have travelled as yesterday a little west of south near the base of the mountain, over numerous streams coming from it that cross the road. Grazing in this valley or upon the east side of the mountain or road extensive & good. No timber to be seen except some scrubby cedar two or three miles up the mountain. Have travelled 24 or 26 miles today and are camping for the night to the left of the road near the heads of two fine springs. Passed nine dead horses.

21st day of August. Wednesday. We have reached the pass or opening in the mountain, through which the road runs west, after travelling ten miles nearly south. It is twelve or fourteen miles over this mountain to the stream or creek in the next valley, which lays nearly north and south. There is neither water nor grass after we leave the bottom of one valley until we reach the creek of the other. The road here follows down this stream north, upon both sides of which there is plenty of grass, but very *coarse*. At the crossing we see several notices stuck up by those who have passed, warning the emigrants to watch their stock saying that the indians have stolen from them horses, mules, and oxen. One billet informs us that 4 packers were killed by the Indians. Road over this mountain better than could be expected. Four dead oxen and one horse. Have camped for the night about eight miles down the stream from where we first struck it; making our drive today about thirty miles. Road down the valley good. No snow to be seen this side of the mountain. Nights frosty and very cold; days hot.

22nd day of August. Thursday. Started early, and a mile from the camp I found and Indian *bleaching* by the side of the road, a *dead* instead of a *living* evidence that some emigrant had done him justice. A few rods further I saw a notice upon a paper stuck into a stick written the 19th by General Rains, who says they laid over that day to bury two men they found were killed by the Indians. Near them they also found a dead Indian. Still further on they found a man shot through the heart and his tongue cut out!! This side of where we crossed the creek, there came to us four men from the mountains which they crossed five or six miles north of the road. In their travel over they saw some Indians and fresh tracks of horses, mule, and oxen. We learn that 20 men from Clay Co. Illis. intend to go up in the mountains for stolen stock and to shoot every Indian they see. This is the watch word with nearly all the emigrants. Travelled today E. of north upon the side of a fine running stream of water, with the best of grass upon the flatts which are extensive.

23rd day of August. Friday. We are staying here today for the benefit of the teams. Some of those who went out yesterday have returned with 6 horses and have killed several Indians.

24th day of August. Saturday. Morning, and effort was made this morning to go up in the mountains where we saw an Indian fire. But Doctors Palmer from Michigan and Woodruff from Wisconsin opposed the expedition; nothwithstanding they saw the day before two bodies, or men unburied, and bleaching in the sun; with their tongues cut out by the Indians. Yet this painful, shocking sight, failed to produce any disposition in the minds of these two men, to take just vengeance upon these wicked demons. They had not a single tear of sympathy to shed over these murdered men, their hearts were indifferent, and cold, careless about the lives of others if their *noble selves* are *safe*. Such men are a disgrace to the world. After some debate we abandoned the idea of chastising the Indians and went on. We have passed over a large stream coming in from the mountains upon the right of us. The road follows down this stream a little west of north, along which the grass is good. Travelled today about twenty miles. Passed 3 dead horses and two mules.

25th day of August. Sunday morning. Looks very much like rain. I shot a creature, or beef for Mr. Hall that was spring poor; yet it sold readily for 12 and 15 cents a pound. We have travelled about 15 miles down a canyon that is by far the most difficult of any upon the whole road. Crossing, and going down the channell, swimming the oxen, and floating our wagons for ¼ of a mile at a time. We have got all our clothing and provisions wet. Passed two dead horses, 2 oxen, & one mule. And are camping at the mouth of the canyon, in sight of some road north of us, which evidently intersects this, There is not an emigrant upon this road that knows where he is. Some fancy we are near the base of the lofty Sierra Mountains; while the more sensible, or better informed, are forced unwillingly to believe that we are just coming upon the valley of St. Maries River; rising of four hundred miles from Sacramento City. Many of the emigrants entirely without provisions, and the balance with very limited stocks. Such a situation could not fail to create fears in the minds of us all.

26th day of August. Monday. Noon. We have been detained here drying our clothes, and provisions; while Hall was hunting one of his oxen which he could not find. This P.M. we have reached the north road from Salt Lake and learn from emigrants who taken that route that we are still 430 or 40 miles from Sacramento. They have been two weeks last Friday coming from the Lake, we three. I am sensible that we have travelled since leaving the valley not less than 500 miles. This makes our road a *cut on* instead of a *cut off*. The people who have taken this route are nearly starving; they expected when leaving the valley to get through in 20 or at most 25 days, whereas we are 440 miles from the city and have been 25 days upon the road. I was amused at the dissappointment of Dr. Lull of Madison, as well as some others, who camped with us night before last; where we examined the maps from which he concluded we were opposite Carsons Lake within 150 miles of the city. I told him that the ranges of mountains that we crossed and the stream that we followed down, plainly marked upon the map were satisfactory evidence that we were on the east side of the great basin formed by the St. Maries River, far above the sink. I saw the Doctor again last night who acknowledged the *Corn*, greatly against his wishes, and now believes that we are far from the end of our journey. Those who have taken the northern route found grass good and water generally plenty, numerous *lies* to the contrary notwithstanding. A man here in shooting at a deer this P.M. bursted his gun tearing his hand and wrist badly which has been amputated by Dr. Lull.

27th day of August. Tuesday. Started early, and have crossed the St. Maries River three times this forenoon. Fording good. The stream is about four rods wide, and two feet deep, clear and rapid. Roads good, but some hilly. This narrow valley is enclosed with rocky bluffs, and covered with dry grass. We have taken our *grub* upon the bank of the river; which we leave here to our left and pass over points or spurs of the mountain a distance of 17 miles where we again come to the river. This distance from where we camped to the small stream or mountain is 14 miles. Passed today two graves. The first upon the left hand side of the road, was Jas. W. Gray of Penn. died August the 14th. Our road over the mountains was hilly, dreadful dusty, stoney, and remarkably cut up, tiresome to the teams and hard on the wagons. Five or 6 miles from the base of the mountains we came to a small spring 50 or 60 rods to the left of the road, which soon sinks, making a dry stoney channel that crosses the road. Here we took our suppers, watered the teams, and remained until the moon rose about ten O'clock when we moved on. We reached the river an hour or so before day. Yoked the oxen, and straightened ourselves out to rest. Passed during the day and night 3 dead horses, one mule and 6 oxen. Since we have come up this stream, we have found the nights more comfortable and warm and we have suffered less from the scorching heat of the sun. Some distance back a Doctor from Montana gave $56 for 112 lbs. flour. A few pounds have been sold here for $1.50 per lb.

28th day of August. Wednesday. We are told by Mormons returning form the mines to Salt Lake that at the sink of this river there is flour to be bought for $1.50 to $2.00 per lb., and beef from 12½¢ to 50¢ per lb. It is brought from California. Grazing where we have struck the river very poor. We have crossed and are going down upon the south side. A packer informs that some of Captain Oliver's train have had their teams stolen by the Indians. This train was laying by on account of sickness; and upon the 21st we passed them. After travelling four miles down the river for grass we camped, where we will remain during the day. The most of emigrants now live upon the flesh, or I should say, the *soup and muscles* of such cattle as fail and are unable to travel. The best stock are all wanted for the teams and packing. Just above us an emigrant is camping who had lost six horses, stolen by the Indians.

29th day of August. Thursday. I am told this morning that a mare was led past our camp last evening that resembles my little bay filly. I started to overtake her, which I expected to do by noon. But at 10 O'clock after travelling ten miles I came to the point of a mountainover which the left hand fork of the road passed, and the right crossed the river. I saw the left hand road inclined from a southwest direction; this road I took. Horatio, who was some distance behind, took the same road, but soon found that Mr. Hall had taken crossed the river; he then directed his course across the bottom for the river which he forded; and after travelling eight of ten miles back reached the camp long after dark. The flatts of the river after leaving this stoney point enlarged to a distance of 25 or 30 miles, and covered with numerous roads. I was soon convinced that unless I went in advance of the teams until I came to some place where the mountains would approach each other, and unite these roads and then wait, I would hardly find them until we reached the sink, 189 miles from where we parted. I travelled until 2 or 3 O'clock in the morning without seeing a human being, at the same time suffering from much thirst and fatigue. I threw myself upon the ground two or three times to sleep, and rest my wearied limbs. But alas!! I was soon compelled to travel or perish with the cold, as I had neither coat nor vest. A little before day I came to some wagons and took lodging with two men sleeping upon the ground for an hour or two. Five of 6 miles from where Mr. Hall crossed the river, he saw the grave of Ephraim Boles of Keokuk County, Iowa; who was killed by the Indians on the 19th in a fight up a dry creek ten miles to the right of the road. Five or six miles from this place, where Mr. Hall camped for the night there came running from the mountains four young oxen and one cow. He supposed they had escaped from the Indians and drove them on with his stock. Horatio got in at 9 or 10 O'clock at night.

30th day of August. Friday. I started early. The river and road here make an acute angle bearing north east for 10 or 12 miles where they again take a west direction. While travelling over this distance a man from Morgan County, Illis. went down to the river from the road where he met several Indians who instantly raised the yell and pursued him. They fired several arrows at him, but none took effect. When he reached the train several returned with him to the place where the Indians were, but they had safely concealed themselves in the tall grass and densely thick willow bushes, so that not one could be found. A little before sundown I saw a number of teams bear down to the river which I followed. They struck their tents close to the river and opposite two large mountains, which separate the north road from the river. To the left of these mountains which can be seen a far distance, the emigrants should cross the river, and keep upon the south side to the sink. This would save 30 or 40 miles travel. After some pleading I have got my supper with a promise for lodging and breakfast from one Chappel who is moving with his family from Montana to California. Some of his train went to the river for water, a distance of 12 miles or more, who came to where 40 or 50 Indians were herding some fine horses, which have been stolen from emigrants as they are shod and have the marks of the harness and saddles upon them. The Indians gave a yell, mounted their horses, and came up with their bows and arrows over their heads, their sign of friendship, saying, *how do do*. The eight men got their water and were not molested.

31st day of August. Saturday morning. I asked Mr. Chappel my bill, which he said was $1. I gave it to him with many thanks. There was an ox killed here last night for the benefit of the starving emigrants. The price was from 15 to 37½¢ per lb. A few pounds of bacon was sold here for $1 per lb. I heard a man offer $4 a lb. for flour but failed to get any. The atmosphere upon the valley is dark and humid like our Indian summer. After travelling 13 or 15 miles I came to a stoney point in the mountain, over which the road passes near the river, leaving a valley upon the other, or north side of the river of about four miles. I reached this stoney point at 9 O'clock A.M. where I watched for the team until dark. This elevated point of mountain gave me a good view of the opposite side. Just below me a train camped which I recognised to be the same that I stayed with last night. I went to Mr. Chappell, my former land lord for supper lodging and breakfast again. He told me that he was afraid to spare a mouthful more. I then went to all the other who preferred keeping what little provisions they had to any amount of money. Some time after dark Mr. Platt with two others drove up, they invited me into their tent, gave me my supper and breakfast.

1st day of September. Sunday. In the morning I offered them pay which they refused. At 2 O'clock P.M. I saw Hall's team moving upon the opposite, or north side of the river. I swam over, and in a short time overtook them. I shall now stay with the team until we eat up our provisions; which will not be long, as we have to divide with those who are travelling with us. The road here leaves the river to the left and passes over some uneven mountains and sandy ground for 8 or ten miles where it again strickes the river. At this place we camp for the night. Good grazing; and willows for wood. Passed today four dead horses and six oxen.

2nd day of Sept. Monday. From this place the road bears east of north, and rounds a slough ten of twelve miles from the river, until it strikes the base of the mountains where the road bears east of north, and rounds a slough ten of twelve miles from the river, until it strikes the base of the mountains where the road bears south of west close to the river. The mountains here are four or five miles apart although the grazing flatts are only from one to two miles wide. The other distance, or benches of the mountains are covered with prickly fern, grease wood and wild sage. Passed today upon our left the grave of George W. Davis of Iowa, aged thirty three years. Passed 14 dead horses and twenty-two head of cattle. Some portions of the flatts here are covered with Saleratus in such quantities that it can be gathered up perfectly white, clean from the earth. In many places the grass is formed into alchalie pyramids as hard as ice, and so sharp that my feet are made sore in thick soled boots in travelling over it. It is the grass upon such ground that produces so much mortality among the stock. We have driven today about twenty five miles. Some mule packers came up after we had camped for the night who told us that they found a man in the river who had been murdered by the Indians. While they were burying the man a large gang of Indians appeared and were soon out of sight again. Many are forced from necessity to *hunt, fish, and kill* frogs to avoid starvation. This strangling alone over the mountains is extremely dangerous. For safety the emigrants are now gathering in large trains.

3rd day of September. Tuesday morning. Our camp was thrown into confusion about 10 O'clock at night by the running away of the *cattle*, breaking loose of the horses and mules. They were *stampeded*, as such scares are called. Mr. Hall put upon one of his oxen a bell that I bought in Galena; this enabled them to determine the direction they took, while pursuing them in the night. And this morning, the sun two hours high they were all brought back, harnessed and yoked, and we again on our way to the gold world, smiling and rejoicing over our good luck. The direction of the road here is southwest, and for 14 miles is sandy. Passed today ten dead horses, eighteen oxen, and four mules.

4th day of September. Wednesday. We have camped within a mile of an elevated sand hill or mountain upon the bank of the river. In starting we go a little north of west towards this mountain and as we approached it the road inclined to our left or south leaving the mountain some distance to the right of the road. Opposite this place we struck a north bend in the river which we followed down for 12 miles, when we ascended a sand plane or elevated table leaving the river to our left. This plane might properly be called a desert, as there is no vegetation upon its surface; a distance of fourteen miles, when we again come to the river and are camping for the night. Reached the river at 10 O'clock at night. The banks of the river down which we have travelled today are from 50 to 150 feet high. Passed today two graves, 23 dead horses, 5 mules, and 17 oxen. Grazing along the river bottoms very poor as they are growing quite narrow. The road over this plane is south west.

5th day of September. Thursday. Standing at our fire this morning, I counted 26 dead horses, 6 mules, and 17 oxen. The road from our camp goes S.W. but soon turns west, making a large elbow, caused by the river, which curves south. We have travelled today 14 miles and have camped upon the bank of the river. We are told by Mormons, returning to Salt Lake, that there is no grass from here to the big meadow, a distance of 50 miles. At this meadow which is 25 miles above the sink of the river, the emigrants cut hay for the teams over the 40 mile desert. Road toady very dusty although not so sandy and heavy as that we passed yesterday. After leaving this morning, I counted close to the road, 19 dead horses, 5 mules, and 13 oxen, making in all one day 45 horses, 11 mules and 30 creatures. One grave.

6th day of September. Friday. The road inclines S.W. from our camp for the distance of five miles where we strike a bench in the mountains, over which the road passes, due south, leaving the river, which takes a large bend to our left or east. We have travelled upon the bench or desert 12 miles when we turned to our left from the main road for the river where we are taking a scanty dinner among the sand hills without any grazing for the stock. In the P.M. we have continued our journey over this table or desert for 5 or 6 miles or more. We have turned off the main road from a south to an east direction for the river where we are camping for the night upon an elevated bank of the stream without anything but grease wood upon its surface. What water we require is carried up this mountain one half mile or more. No grazing of any account for the horses. Passed 18 dead horses, 8 mules, and 14 oxen. Two men from Platville called at our tent who were without provisions and money.

7th day of September. Saturday. The river along here is

hemmed in with high broken clay bluffs, leaving no valley. The direction of the road today has been regularly south over this desert bench of the mountain for 14 miles where we have again come to the river. As usual, not much of any grazing. Here we take an *allowance* dinner. Have passed one grave, another close to the bank of the river. This A.M. we have passed 42 dead horses and 23 oxen. Roads very badly cut up and very dusty. In the P.M. after raising the steep bank of the river I counted 13 dead and 7 oxen upon a small piece of bottom ground to our left. From this place until dark I counted 26 more dead horses and 17 cattle. We travelled until 2 O'clock in the morning, and from our *nasal* knowledge of numbers have passed about as many more dead creatures. Making a probable number of 162 dead horses, and 94 oxen. As near as we can calculate the distance, we have travelled since dinner 26 miles; making during the day and night 40 miles; this places us far down in the *Big Meadow* where we are trying to get some grass for the desert which is 15 miles further on. Grass here is very poor, and this sandy bottom does not deserve the name of meadow. We now witness such suffering from fatigue, want of food, from thirst and cold during the nights. Before leaving home I anticipated all the unpleasant, painful sights that I have yet seen. But then *I* little expected to come in for so large a share of suffering as has been our lot to bear!! The water here is unfit for use. We have felt *fatigue* in all its *forms*, *hunger* with all its knawings, and *thirst* with its parched, burning influences. Indeed we are now getting a naked view of the *elephant*; he is an awful animal, breathing out pain and misery in all ways and forms. This meadow has an immense number of dead creatures scattered over it. If our horses had not been stolen we would have probably been in California by this time, and have avoided much danger to which we are now exposed.

8th day of September. Sunday. Another holy day has been

been spent by us making hay, killing an ox, jerking the meat and resting the teams for the desert. A large number of Indians are camping 15 or 20 rods above us: from these we learn that 2 or 300 are living 4 or 5 miles below us. They have travelled from one camp to another, with guns, upon some fine American horses. These Indians are short and thick, rather corpulent. They are very well clad with American clothing. They call themselves Piutes.

9th day of September. Monday. We are yet drying and jerking meat!!! Such universal time of *jerking* I have ever seen before. We have paid $6 for what you could put in a good sized bake kettle.

10th day of September. Tuesday. The Indians are again at our camp. They make signs to us, by which we learn that several of their tribe are sick. Some of them were here last night and gave Doctor Harris two horses for four boxes of pills. They want us to go down this morning and see their sick. Doctor Harris, myself, and two others belonging to his company have concluded to accompany the Indians, 6 in number, to their camps down the river. In four miles travel we found them in squads among the wild corn, which grows upon this bottom as thick as it can and stands from 6 to 10 feet high. Upon this unhealthy marsh, surrounded with this corn and grass we found them. Several died yesterday, some are now dying, and a large number will die before morning with the congestive fever. The Doctor and myself began to deal out medicine to those that we thought had any chances to live; which they took with astonishing eagerness. They took us from squad to squad until noon, when we became sensible that we had got a whole days job upon our hands. We told them, by signs, to bring up their horses, that we must have one a piece, to enable us to overtake our teams. They sent out for the horses, some of them were poor in flesh which we refused to take. After some picking we got one horse a piece, when Doctor

Harris and I began to administer medicine to the rest of the sick which we did as quick as possible, then mounted our nags, and took a west direction for our company where we overtook them just at dark taking supper. In our ride by the side of Humbolt Lake which was to our left, we passed six graves, 37 dead horses, 42 dead oxen, and five mules. We started upon this desert about one hour after dark; drove ten or 12 miles, and stayed until daylight. From *nasal* evidence we were continually passing dead stock.

11th day of September. Wednesday. This days travel has presented to us sights that are truly painful. Horses, cattle, and mules, lie here in heaps. Wagons, carriages, carts, harnesses, clothing, and furniture enclose the road, and in many places literally block it up. Up to this desert I have counted all the dead stock and have attempted it here, but I counted one hundred and eighty-three dead creatures and did not get out of the sight of the first. So I gave it up, and when we got over, I consulted several persons about the number, who estimated the whole from 4 to 6 thousand. This number 5000 at an estimate of $40 a head which is a low one, amounts to $200,000. Twelve hundred wagons at $60 a piece makes $72,000. One thousand harnesses at $30 a set makes $20,000. This amount added to the value of ox yokes, guns, clothing, household stuff etc., that cover this dreary desert as far as the eye can see will enlarge this sum to the startling amount of $300,000 at least; and perhaps half a million. This is a large sum to be left upon the short distance of forty miles. The road in some places was completely blocked up with wagons and dead stock, some unharnessed and still to the wagons. The last twelve miles are very sandy, so that an empty buggy is a full load for a team. The road is some sandy over the whole distance, and the wind covers up to a considerable extent the dead creatures; so that the stink is made by this covering less, although now almost intolerable. The direction of the road across is due west. Seven miles from the west side of this desert we came to a water and liquor establishment. Water sold at $1 per gallon, whiskey at 50¢ a glass, and lemonade at 50¢ a glass. We go over just at dark and are camped upon the east bank of Carsons River. This is a fine stream of good water about eight rods wide and from four to six feet deep. No grazing less than 15 or 18 miles from here down the river, and 30 miles up, upon our road. We find grass here for sale from 12½¢ to 25¢ a bundle. I bought eight for my pony and carried them under my arm to the camp a quarter of a mile. Doctor Harris bought 100 bundles for his creatures, 13 head, for which he gave $12.50. It made only a few mouthfulls a piece for them. Eight graves over the desert from cholera.

12th day of September. Thursday. We are now among the canvas houses or trading posts buying provisions. Flour sold here a few days ago for $2.50 per lb. Pickled pork for $2.50. Cheese, coffee, sugar, & tea the same. Today I bought 15 lbs. flour for $6 or 40¢ a lb, 8¾ lbs. pickled pork for 87½¢ per lb; and two pounds of Mexican sugar for one dollar. Molasses they sell for $3.50 a pint. Cheese sold today at $1.50 per lb. and coffee at 75¢ a lb. This sudden fall in prices is caused by the cholera here, and the emigration taking a northern, or what is called the Truckey route. Grass is said to be plentiful upon this road; which leaves the one we have taken the other side of the desert. I am told by one of the traders here that no less than fifty persons have died here of the cholera since yesterday morning. Three brothers from Montana by the name of Lewis made some arrangements with Mr. Hall to carry some freight across the desert for them. They had one horse among the three, which one of them who had not recovered from the mountain fever intended to ride. The horses gave out the other side of the desert. Mr. Hall took Lewis into his buggy until he came to this 12 miles of heavy sandy road where he got out and walked a mile, when he got a chance on horseback. In this way he got separated from us. In the morning before day, two men came to our camp and told us that he was out a part of the night, and in crawling around fell, near an abandoned tent into this he found some old clothes with which he covered himself and was then dying. The two brothers got there as he was breathing his last. We are camping for the night upon the bank of the river four miles above where we first struck it. Turned our cattle into the willows to brouse until morning. In two miles travel we leave the river and pass over a desert point of the mountain for 15 miles before we come to it again. In the short distance of four miles which we have travelled we have passed fourteen graves and an immense quantity of dead stock. The road runs up the river north of west.

13th day of September. Friday. In about three miles from our camp we came to a rough stoney bench in the mountain over which the road runs for 14 miles where we again come to the river. From here the road passes up near the river upon the bottom or flatts which have furnished fine grass, but the large quantity of stock have consumed it all. In this distance and around our camp I counted 26 graves, 53 dead horses, 79 dead creatures and 11 mules. In my opinion the most of the deaths since we struck the desert have been caused by the unhealthy air inhaled from such a mass of putrifying stinking matter, living exclusively upon very bad beef and too free use of the very worst of water which the emigrants have been compelled to use for 60 miles the other side of the sink until they reached Carsons River. All these deaths are from cholera. Our travel today has been about 21 miles.

14th day of September. Saturday. We have stayed here all day to rest the teams and give them an opportunity of filling up their lank bodies once more. It might with truth be called an entire *desert* from the sink to this valley. Where we are camped there are two flour and *grocery* establishments. I have bought 10 lbs. of more flour at 25¢ a lb; also a ham of 12 lbs. for 75¢ a lb. from

Bryant an emigrant from Montana. Horatio and I are both some unwell, I think from eating *jerked beef* and the other causes already mentioned.

15th day of September. Sunday. Soon after leaving our camp we raised a high sandy hill or table elevation of the mountain over which we travelled for 13 miles; leaving the river to our left. This distance is an entire bed of sand. From where we struck the river to where we are camped upon its banks the distance is six to eight miles. This valley from one bench of the mountain to the other is from one to two miles wide, and furnishes good grass, plenty of small, dry *elm* and willow, and any quantity of cotton wood. The water is not cold but pleasant tasted. We can see from here the tall, white, snowy heads of the *Siera*. Passed today 13 graves, 39 dead horses, and 43 oxen and cows. We learn from a trader, just from the desert, that on Thursday, the day we left there, or during the day and night twenty died with cholera. He also told us that flour was sold there that evening and the next morning for 15¢ per lb.

P.S.N.B. Please send your letter to me directed to Sacramento City. There are persons, called 'Express Letter Carriers' in all the mines in California. By them I can get your letters better from Sacramento than any other place.

1850-16th day of September. Monday. Three miles from our camping place we crossed the stream upon the south side. The fording about two feet deep and from four to six rods wide with smoothe sandy bottom. The general direction of the river today has been a little south of west. The road in some places very sandy, and others rocky. We are camping three or four miles east of where the road again crosses the river. Grazing very poor. The banks of the stream here are elevated, near each other, and covered with black burned stone and other evidence of volcanic disturbances. Cottonwood growing along the stream. Passed two graves, 13 dead horses, 20 oxen and cows, and 2 mules.

17th day of September. Tuesday. Soon after crossing the

river which is good fording, we raised an elevated stoney bench of the mountain leaving the river to our left. The road upon this bench goes little south of west. From the crossing of the river to it again a distance of 12 miles is very sandy, and the balance dreadful stoney. Three and a half miles before we come to the river the road passes over a ridge of beautiful white plaster, a large quantity of which the wagons have ground up. About three miles from where we struck the river we have camped in a small valley of good grass and plenty of water. No wood convenient except sage brush. After passing into this valley we came to a hot spring of water at the base of a mountain putting into the valley on our left. The water is so hot that a person cannot bear their hand in it. This P.M. we met a relief train, with flour, who were ordered by the citizens of San Francisco to cross the desert and there remain as long as the season would permit. Passed today 3 graves, 34 dead horses, 28 oxen or cows, and 7 mules.

18th day of September. Wednesday. We have driven six or seven miles today, which takes us into what is called 'Carson's Valley.' Here is the last good grazing we find upon the route. The grand lofty mountain upon our right is timbered with pine and in some places covered with snow.

19th day of September. Thursday. Our direction today up this valley has been nearly south, passing several fine gravelly streams of water dashing down from the lofty Sierra through an immense forest of majestic pine, fir, and hemlock timber. If I had 500 acres of this timber at our mill the *gold* in *California* would not be disturbed by me. This is a beautiful valley 24 miles long and about 20 or 25 miles wide with an abundance of grass, which is cut and cured by the emigrants for their teams over the mountains. There are some sloughs covered with tall willows, where horses and cattle are difficult to find; sometimes lost, and afterwards picked up by the Indians. I bought this P.M. 23 lbs. of flour at 15¢ per lb. Passed today a relief station, or a place where *beggars* can get what *flour* they can eat. 27 dead horses, 19 oxen, and 2 mules. Made today about 15 miles.

20th day of September. Friday. We have made a distance of five or six miles today. I think Mr. Hall from his short drives intends to fit his cattle for the California beef market. An immense number of streams put out from the base of the mountains to our left. Two men this P.M. in our camp of diarrhea. One was about four rods from our wagon. Soon after we stopped I heard him groaning; about this time one of his company came over who told me that the man had the *hypoc*. While he was talking with me the groaning ceased, and the poor neglected sufferer breathed his last. I went over, but he had taken his leave of this world, among his unfeeling companions, who changed his clothes & soon placed him in his final home.

21st day of September. Saturday. We started about one or two O'clock in the morning, travelled about ten miles south, up the gradual kanyon at the head of this valley. Carson's River is formed by streams putting out from this, and other defiles or kanyons making into the valley. The road takes a west direction through an immense of timber and massive rocks that continually echo and re-echo the beautiful sweet songs of melody, made by the player Mr. *Wind* through the branches or strings of these pine *organs* of nature. Here is *nature's theater*, upon a grand scale I can assure you; the mossy *drapery* or hangings have been beautifully arranged by nature in her playful moments. It is the most glorious, magnificent sight that I have ever seen. Passed today 23 dead horses, 14 creatures and five mules. We are camped for the night four miles up this kanyon. A heavy steady rain this P.M. and night. This is no road for wagons, passing over stone from two to five feet through.

22nd day of September. Sunday. No human being can pass through this awfully formed part of creation without entaining thoughts peculiarly pleasant, instructive and sublime. Here, far above us, as it were, in the second house of heaven we behold the tall pine and fir tree clothed in beautiful rich green singing sweet plaintive songs and praises to God; while the white hoary headed spires and monuments of granite mingle with, and hold communion among them, emulated by a jealous ambition, are equally busy in carrying up the echoed sounds the blasphemes, the dams and curses of the insignificant emigrant *far* very *far* below them. Here I am forcibly struck by the following beautuful passage of scripture, "Lord, what is *man* that *Thou* are mindful of *him* or the *son* of man that *Thou* should regard *him*." A mere note in the immensity of thy works!!! We are now through this bad canyon into what is called the ten mile valley. This valley is an area of about 200 acres gradually inclining to the south, W. & north surrounded by lofty mountains, some of them well covered with snow. Four or five miles from the road, to the north, upon the side of the mountain, there is good grass. From the mouth of the canyon to this valley I think the distance is about ten miles. Here is a large quantity of water gathered from the mountains, and dashed down the gulch that we have come up into Carson's River; making one of the most important tributaries of that stream. Passed one grave, 38 dead horses, 13 mules, and 7 cattle. In this valley we camp for the night.

23rd day of September. Monday. The road from our camp takes a south west direction, raising an elevated bench of the mountain, upon which we are camped, four miles from its base, and five or six miles from where we stayed last night. This piece of road is very good with the exception of the crossings of two streams, which are very mirey; and almost covered with swamped stock. It has misted and rained all day, and at night a violent tempest came up that gave us torrents of cold rain, and clothed the high peaks with a mantle of snow. I have this morning put myself in flannels and thick clothing, which feel very comfortable.

23rd day of Sept., continued. Passed today the grave of an infant four days old. This delicate germ of humanity found this too rough and forbidding a country to suit its infant tenderness, therefore took its exit to a smoother and happier world. Passed today 24 dead horses, 7 mules, and 15 cattle. Grazing here, among the stream from the mountains although dry yet plenty.

24th day of Sept. Tuesday. Rainy and very cold this morning. Two of the oxen got into the willows in the morning during the rain and could not be found until near night which detained us in the cold, bleak, disagreeable place the entire day. Mr. Hall sometimes manifests a total indifference whether he ever gets through or not. I have grown heart sick of the journey.

25th day of Sept. Wednesday. We have again got ourselves under way, with our faces towards the Sierra berges. The road runs from this station a little south of west, and in a mile or so inclines to the west, raising rapidly the smoothe base of the mountain for four miles; where we came to a very rapid rise of one mile so steep that the best yoke of oxen could not draw up an empty buggy. This piece of road is rendered particularly difficult on account of the large smoothe, flat faces, inclined rocks that obstruct the road. When a creature steps upon these and makes and effort to pull he falls down as quick as if he were upon an inclined body of smoothe ice. With the exception of two short turns in the road any number of teams can be used before the same wagon. Indeed I saw fourteen yoke of oxen before a common wagon with eight or ten hundred pounds on. Such hollowing, swearing, and cracking of *Mo. Hemp* I have never heard before, and never wish to again. This wagon with twelve others belong to Bryant of St. Louis, Mo. They are mostly loaded with ham, which he sells at 75¢ per lb. These teams got to the foot of the mountain before us, and were all day getting the one mile on top of the mountain; where they were forced to tie their stock to the trees and wait until daylight. Consequently we camped between Red Lake, a long body of water, and the foot of the mountain. The wind from the snowy hills, whistling through the pine and fir, gave us a cold atmosphere and plenty of *music*. The place is literally covered with dead stock. I have counted today 69 horses, 23 mules, and 8 oxen. One grave.

26th day of Sept. Thursday. Started early; put five yoke before one buggy, drove that up and then returned for the other. From here we descended the mountain which we found in some places dreadful steep and stoney. We are camping at a fine lake of water at the base of another mountain, up which the road goes; taking a south direction from the N.E. past the lake. Two or three miles north of this lake up in the mountains we have driven our stock, where we find very good feed. At the lake the grass is very short. Our travel today has been about five miles; passing 15 dead horses, 7 mules, and three oxen.

27th day of Sept. Friday. Today we have passed over another mountain whose tops have taken a peculiar fancy to the upper world, and a hatred to the earth beneath; which is occasionally pelted with immense rocks, breaking from the mountain, making heavy California thunder and lightening in their angry, warlike descent upon whatever is below. It is impossible to describe the bad condition of the road over these mountains. The distance over this one is about four or five miles; in some places very steep, but clear from these immense rocks over which we have come. Before reaching the summit of this mountain we passed over the edge of a snow bank twenty feet deep, which has probably been here for centuries. From the top of this mountain we have travelled two or three miles, ascending and descending very steep, rocky places. We are camping among the fir trees and rocks which try to rival each other in height. A hard country for stock. Passed today 56 dead horses, 9 mules, and 7 oxen. Old *Sol* is very kind in spreading his comforts over these dreary places.

28th day of Sept. Saturday. We have descended this mountain to rock valley. A smoothe spot of ground to the right of the road. Here we find a man from St. Louis with others hunting the woods for his wife, who was seen yesterday evening taking a path, which she thought, probably, would save travel and come into the main road again. This path is an Indian track leading from the road into mountains. From this place we ascend and descend a succession of mountain ranges to another smaller valley and a fine spring of water close to the road. The stream crosses the road. This place is called *"Tragedy Springs"* from a murder committed upon Cox, Blunt, and another man in 1848. We are camping four miles after passing these springs at the top of a gulch to our left, down which we have turned the creatures for water and what grass they can find. The road today has not been as stoney as we found it for the few days past. Hemlock, fir, and pine timber of a large size in abundance. Passed today three graves, 65 dead horses, 49 oxen, and 9 mules. A pleasant day.

29th day of Sept. Sunday. We have struck our tent in 'Leek Spring Valley'; so called from the quantity of *Galick,* called leeks growing upon the flatts of the stream made by a fine spring coming out of the ground close to the road as we come to the foot of the hill upon our left. There are two trading posts here. These traders call the distance from here to Weaver & Hangtown 50 miles and from these towns to Sacramento City fifty miles. From these traders I bought some fine California beef steak for 30¢ per lb., and some pickled pork with bone for 63¢ per lb. Some two and a half miles before we came to this valley we passed down Iron Hill quite a steep long pitch. This hill has its name from the iron taste which the water has and other appearances about the mountains. The road today has been hilly, but not very stoney, timber the same as yesterday. Passed 27 dead horses, 23 oxen and cows, and three mules. Passed four graves about eight miles west of Tragedy Springs to the right of the road. One of them contained M. D. McCan of Mineral Point, Wis. died July 27th, 50. The others were from Iowa State. The lost woman came to this trading post today; much injured by falling over precipices and rocks while travelling in the night.

30th day of Sept. Monday. I have thrown my blanket robe upon my pony with little provisions and started for Weavertown in advance of the teams. Horatio drives the team in my place. The road to Deep Valley, a distance of about twelve miles from Leek Springs, runs nearly west and is very rough; descending long steep and stoney hills. Here there is a small stream, the Mormons call a river. No appearance of grass. From a trading post here I bought two pounds of rotten stuff, which they call *hay* for my horse, for 50¢. Sixteen miles from Deep Valley I am lodging with Mr. Watson who generously furnishes me the use of a tent an excellent cup of coffee for my breakfast. The road has been very stoney, passing over many long steep hills, both ascending and descending. The few scattering oak trees, anywhere near the road, are all cut down for brouse, to sustain the stock. Here I have paid 40¢ a pound for 4 lbs. of barley for my pony ⁵ $1.60. The weather since we passed the summit on the 27th has been pleasant. Passed today 55 dead horses, 47 oxen and 13 mules. Nothing like grass to be found upon this hogs among the pine and hemlock.

1st day of October. Tuesday. A delightful morning. I have travelled today nineteen miles and have camped one mile from Weaver. The first ten miles of my journey today was west; the road hilly and rough. About four miles from where I camped the road forks; the left hand goes to Weaver, the right to Hangtown. Ten miles from Weaver the descends a long hill and drops upon a pleasant narrow valley. Here the grand and lofty scenery gives way to a lower and more familiar kind. The while oak & the live oak with their low mossy tops extending far over the trunk afford a beautiful as well as comfortable shade to the wearied traveller.

The road for the last ten miles is gradually descending, smoothe, and fine, passing down this narrow valley, which has had from appearance some grass. Emigrants cut a great many oaks for browse to keep their creatures from starving. Passed today two graves, twenty-five horses, thirteen mules, and twenty oxen.

2nd day of October. Wednesday. Warm and pleasant. I am now in Ringgold. It is located in two ravines, one running east and the other north and south. Has a population of eight or ten hundred. Ninety houses or huts scattered among bushes, oak, and pine trees. I have seen but six or eight females in town. The style of building is peculiar to the place. Holes are dug the size of the building in which small pine posts are placed; pine poles are spotted and placed upon the tops of these for plates and end pieces. The house is then enclosed with pine shecks, or shingles, split with the faces six inches wide, and from three to five feet long, and covered with the same. Some of the miners here are making some money beyond their expenses. But a very large majority only clear their board. Weaver town is nearly one mile north of Ringgold upon the north and south ravine; is in size, architecture, style, manners, fashions, and population similar to it. I. T. Lathrop has opened a grocery and provision store here. From him I learn that Major Gray is in or near Georgetown, mining and has done very well. Hamilton is in the merchantile, grocery, and provision business in Sacramento City. Nic Turner is in Nevada City, Tilley Alex Turner, and the Wasleys are up on the Uba River. Josh and Joe Baileys are in Hangtown; three miles north of this place. He can give me no information about Canoutson, Million, Beams, nor Charley Crellen

3rd day of October. Thursday. I am here yet waiting for the teams. Hay is selling here from 12½¢ to 15¢ per lb. I bought six pounds for my horse yesterday for 75¢; and the same today. Barley is worth 25¢ per lb. The rocker or cradle which is used to wash the dirt from the gold costs $15. It has rockers similar to a cradle. Upon this, there is a square box 20 inches in length and width; 6 inches high in which the dirt is thrown. This box has a sheet iron bottom with a large no. of holes in it to keep the coarse dirt and gravel from passing down with the gold. An apron of board or canvas is placed below this box which carries the water with the small dirt upon the bottom of the cradle at the upper end. A cross bar is placed close and solid upon the bottom of the cradle near the lower end, against this bar the gold lodges while the loose dirt and water passes over it. It is then taken out of the cradle into a tin wash pan and washed by hand. When the dirt is thrown into the box one man with his hand to this cradle rocks and pours water in at the same time. This is the process of washing gold in California. Ames round pointed shovels sell here for $15 a piece. Flour is retailed at 20¢ a pound. Pickled pork at 30¢. Corn meal at 20¢. Dried apples 60¢. Cheese $1.00. Butter $1.00. Peaches 60¢. Muscarada sugar 40¢. Rio coffee 60¢. Teas from $1.00 to $2.00 per lb. White beans 25¢. Peas hulled 40¢. Onions from $1.00 to $2.00 per pound. Rice 30¢. Candles $1.50. Bar soap 25¢. Saleratus 75¢. Salt 20¢. Hams from 50¢ to 60¢. Whiskey $4.00 per gallon. Brandy from $2.50 to $5.00 a gallon. Vinegar $2.00. Molasses $4. Irish and sweet potatoes from 25¢ to 30¢ per lb. Tobacco 75¢ per lb. Boots from $8.00 to $32.00 a pair. Common domestic 25¢ a yard. Calico or prints from 25¢ to 75¢ a yard. I see many *pale* faces here, and numerous dissappointed emigrants selling their teams and putting home with all possible speed. From four to eight are continually in the grave yard digging graves. Whether we are doomed to buffet the waves of adversity and want, or bask in the sunshine of ease and prosperity is left for time alone to tell; and we will patiently wait that result. And trusting to the future we now take farewell of the past. We take *hearty* good *bye* of our *journey*, its dangers, its toils, its suffering and its anxieties with the same feelings that the wandering, weary, wayworn traveller amid the burning sands of Africa would hail the sight of some cool grave where he might turn aside to rest his wearied limbs beneath its refreshing shades, and slake his parched lips with its pure limpid waters.

4th day of October. Friday. We have started this morning for Sacramento City; which is 50 miles from Weaver.

Hangtown or Placerville October the 9th, 1850

My Dear Wife,

We are now in California after suffering all that humanity could bear. Some six hundred miles back we had our horses stolen by the Indians. From then through we had to foot it. We hired our freight carried through. I have seen many letters from Wisconsin that give painfull news about the cholera. This gives me uneasiness about you and the children. I have written to you several times, and in those letters requested you to write to me directing your letters to Sacramento City. I have seen the list of letter remaining in the Sacramento post office quarter ending 30th September 1850. In them I find none for me. I am afraid Constantia that our expectations here are not to be realised. I think that I shall winter at Georgetown 40 miles from Sacramento City. Do write me often my Dear. I feel very uneasy about you. Direct in a large plain hand to that city. (Sacramento) We are well. Horatio is ten miles from Weber where I must go tonight. I will write to you every two weeks giving you my journal through. I am in a very great hurry so good by. My love to you My Dear and the children. Constantia, write to me every two weeks.

Affectionally yours F. M. McDiarmid

Greenwood valley, El Dorado County, California
November the 10th 1850

My dear wife,

In this letter I shall give you a brief history of our journey here, and the prospects of *getting gold.* In the first place it is necessary for me to say that I have mailed for you in Iowa, at Fort Laramie, Salt Lake, and other places upon the road my daily remarks up to Thursday the 13th day of June. I will continue to send you the balance every two weeks, which is as often as the mail steamers leave for the U.S. The first and the 15th of every month. From home up to when we had our horse stolen by the Indians 5 or 600 miles from here nothing of any great importance occured. Henry and Adam Helms had their horses taken from them by the Indians by force. Many others suffered the same loss. From the loss of our horses which were stolen between 2 o.c and daylight; evidently when the watchman was sleeping, our troubles began. Since then we have felt hunger, thirst, fatigue, sore feet, and numb limbs, which we bear with the same feelings that the wandering, weary, wayworn, traveller amid the barren burning sands of Africa would hail the sight of some cool grove, where he might turn aside to rest his weary limbs beneath its refreshing shades; and slake his parched lips with its pure limpid waters.

During this *tedious, painful* journey we enjoyed very good health. For this great blessing bestowed upon us, which was not possessed by thousands of others, we have reason to return God much praise and many thanks. Many person were rattled and rolled in wagons over rough roads when dying. This species of inhumanity or cruelty can be excused by those who know the danger that one or two wagons would be exposed to by the Indians if camped by themselves. And upon the other hand delays could not be made for the want of provisions. Which was, with few exception universally felt by the emigrants to California this year. To save thousands from starving many had to eat the best parts they could pick from the dead putrid stock. Last year the people took too much provisions, this season too little. Now, My Dear, for the prospects of *gold* here. I can sum the appearance up in the following short sentence. If I knew before I left home what I do now of California, I would be enjoying the sweet pleasures of a good *bed* by *your side* and the lively society of my little family, instead of dropping down upon the ground to sleep, and putting myself in all shapes to eat my victuals in *California.* The mass of the emigrants are greatly disappointed. Some of them have become disheartened and with what little gold they find they go to the gambling houses in hopes of making their *pile* that way. But in this they are generally disappointed. Dissipation is the next resort. We have seen wallowing in the ashes at our fire a man by the name of Perkins whose wife and family are at Fairplay urging him to return while he is gambling and drinking his time away here. He with thousands of others will never return to their families. We have built a cabin for the winter 12 feet square in the inside; got the roof on, the door made and on, and mostly chinked. It is 3 or 4 miles west of the valley. The population of the village in this valley is between 10 and 1500. The Indians are our neighbors. They have killed or butchered 15 men the other side of a mining town called Weber. The sheriff of this county made the murder known to the governor who ordered him and Major Kinney to raise 200 men at $8 per day to be paid by the state in state warrants and take them prisoners. The men were raised and they went in pursuit of the Indians. Previous to leaving, the Indians gave the inhabitants notice that the land north of Sacramento City was theirs, and they must leave it in two weeks. When they were overtaken a battle ensued; the Indians routed and some killed. An Englishman who fought the Indians was taken prisoner. From him they learn that 3000 are gathered to get possession of this country. Gibson is with us digging; we three average $3 to $12 a day while we have dug. We got into the first mining town, Weaber, the third or 4th of October. Have spent some time in hunting a place for winter. Have been here about three weeks. It is impossible for us to say whether we can make much during the winter. It costs every thing to live here. We have to pay $20 per 100 for flour. Pork from 25 to 50 cents per lb. Potatoes from 25 to 50¢ per lb. Rise 25¢ per lb. Sugar 40 to 50¢ per lb. Molasses from $3 to 3.50 per gallon. Butter $1.00 per lb. Cheese $1 per lb. Onion $1 per lb. Cabbages, turnips and beets 25¢ per lb. other things in proportion. William S. Hamilton has died some three weeks ago in the City of Sacramento of the bloody flux. I have learned that Charly Crellen was in partnership with him and Mat Chilton in a store in the city and some diggings up the Uba. I have not seen any of them. I have not heard from you and the children since I left home; this gives me great uneasiness; particularly as I learn that the cholera has been very bad there. For a few days from 50 to 80 died a day of the cholera in Sacramento City. Two deaths by the cholera in this valley. Fred and William Underhill, George Biggalow and Ireland tent with us. My Dear do write to me as often as once a month at least. Remember me kindly to my little children whose society I miss very much. I cannot reconcile myself when absent from you and children.

Affectionately yours
F.M. McDiarmid

Georgetown California, January the 11th 1851

My Dear wife,

I am now answering your first and only letter of Sept the 1st

50. It is well filled and affords me much pleasure to know that you and my little children are all well, and the good health that your father and mother enjoy beyond what they have had for the past two years. Your letter my Dear has relieved me of an immense weight of anxiety about *you,* and those little *pledges* of our love; especially as the cholera was again spreading *alarm* and *death* over our country. I sincerely pray that the Lord will never permit this awful scourge to return to our country again. I received your letter the thirtyeth of Nov., and have not been able to answer it sooner for the following reasons. About the 25th of Nov. we left Greenwood valley for Georgetown or what is called Oregon Canyon; about four miles from the town. We had to pack our Rockers, or gold washers, picks, shovels, tents and what cooking fixtures would be necessary upon our backs. Thus we had no more *hands* to *hold,* and no more *strength* to *carry* anything more than what was indispensibly necessary to perform our labour. When travelling here it is customary for people except *gamblers* to go on foot and balance themselves by placing upon their backs between 25 and 100 lbs weight. From this canyon we were driven by rain caving in the shaft that we were sinking. From this place we went north, beyond what is very appropriately called ''Ball mountain'' there sunk a ditch 15 rods long to draw off the water from a favourable looking flatt in a larger gultch that had not been dug upon. Upon this flatt we sunk several holes, from which we only got $12.50. While here for the want of money we lived upon an allowance of bread and meat; and a number of days upon bread alone. Georgetown which was eight or ten mile off and the only place where we could get our *Bread* and meat. I should say that we have had in addition to our bread and meat since we returned, two or three meals of potatoes, and two messes of stewed peaches. But this is expensive *aristocratic* living and beyond our income. Sunday the 12th Horatio and I have just returned from town with 50 lbs of

flour; which cost 15¢ per lb. ⁵ $7.50. One ham 14½ lbs at 40¢ per lb $5.80, and 8 lbs of pickled pork at 25¢ per lb, and 9 lbs of potatoes at 20¢ per lb. This is the cheapest provisions we have bought since we have been in the mines. Making 81½ lbs at $17.10. This purchase has left us only four dollars. My Dear, you say that if I had expressed a wish for you to accompany me here, you would have come cheerfully. Let me hear Constantia what would be your plans, supposing I should think it best to stay 2 or 3 years. The prospect for making a fortune here in the time I have allotted to be from home is very uncertain. It looks as improbable as it would for a man to go upon the dug up hills of Mineral Point or the ground where Hamiltons Steam pump stood, and there dig for *lead*. In 15 or 20 days I shall go up the stream which is called Trinity, about 250 miles north from here or up the south or north branches of the Ubas. If the country up there has been as universally examined as it is here, with the streams damed and turned into canals or races, with the flatts, points, or barrs, putting out from the mountains into the streams all dug up then I shall conclude to go into some trading operation or return home. For the chances are few. Notts, Cagle and Sargent I think have gone up the Trinity. Fred and William Underhill are at work near us. Just living as we are. Russell Baldwin is at work upon a bench or slide upon the canyon above us. He came there soon after he got in, with Chrismus Boardway and Alexander Hammon who came through with him. They will probably remain there during the summer. They fortunately hit upon a rich spot, and suceeded in deceiving all those who make any enquiries by telling them that they only made board, and that they were sorry they ever came to California, while at the same time they were doing well; at all events Alexander Hammon told Gibson that they took out $2200.00 in four days. Chrismus Boardway also told him, as the other, in great *secrecy,* that they panned out at one pannful or wash $500.00. Thus their confidential communications soon became public, for Gibson was first fed upon butter and his tongue is still slippery. In my next letter I will give you a description of the country, the people, their habits, living, etc, etc. Give my love to our little children. I think my Dear that I would recognise the baby sooner from the features and looks of the mother than my own. With the greatest desire, and anxiety to see you, Constantia I can only bid you good by.

While I remain
Yours affectionately
F.M. McDiarmid

Oregon Canyon California January the 19th 1851 Sunday

My Dear wife,

In my last letter I promised to give you some sketches of the country, the climate, the people, their habits, appearances, amusements, living etc. The country from the Macosma river north, to the north fork of the American river, is one continual sucession of elevated mountains, and separated only by deep gulches. These mountains are well covered with beautiful tall white and yellow pine, hemlock of an immense size and length, sedar, common and live oak, of a low, scrubby, shady kind, mansaneta and mountain laurel. The last two are always green. This region of country embraces a number of mining towns. The most important of which are Weaver situated upon the head of Weaver creek. South from four to ten miles towards Sacramento city are Cold Springs, Mud Springs, Log Town, and French Canyon. North is *Hangtown,* called so, from the fact that three men were hung there in the first settlement of the town and one since I have come here. Ten miles north is Coloma upon the south side of the south fork of the American river. Formerly "Sutter Mills" where gold was first discovered in California. North West of this is Greenwood Valley. North of this valley is Georgetown and Georgia Flatts. Upon the middle American river, which resembles in size and the rapid dashing of its waters the Grand river in Qu. Canada, there are Spanish Barr, Big Barr, Volcanic Barr, etc. The north fork has its numerous Barrs. The climate of this part of California is in my opinion very healthy. The weather, so far this winter is delightful. Very mild, and with the exception of three storms, the first one snow, it has been very dry. This part of California produces no vegetation. Scarcely sufficient without waste for the *People*. California can boast of a people whose powers, talents and faculties are not surpassed, *if equalled*, by any people in the world. We can carry upon our backs, up these mountains, and down the gulches, heavier loads than any man east would put upon his horse; and that without puffing! Others can drink more whisky and brandy, and swear with more ease and freedom than any people I have ever seen. In these exercises the Missourians carry off the *Palm*; they are good teachers. Their amusements are drinking swearing and about the gambling tables; especially on Sunday. Their looks are repulsive and forbidding; with long bleached hair, *dirty bodies, shirts*, and *frizzy faces*. The most of the food consumed in California consists of flour, meat and potatoes. The *females* of this state. They form but a small part of the population here, are more fair, can bet at *Monty* at *Faro* and play poker with more beauty skill and wisdom than any ladies in *Chrisendom*. I did not mean to use the word *Chrisendom*; for I am not sure that this is in a christian land. It is in my opinion somewhat doubtful. For the future, in comparison I shall use other words. In the largest gatherings and assemblies these *ladies* unembarrassed can drink *wine*, and *brandy*, and give the most appropriate toasts and sentiments of any women in the *world*. I witnessed in Coloma when news reached there that California was admitted as one of the states in the union one of these female exhibitions. I dismiss this class of people by saying that there are some women here that deserve much praise, although few. In speaking of the weather I should have mentioned that since the last of November the mountains upon our East and North and the Coast ranges upon the West have been continually covered with snow since the last of Nov. South and south west is the only direction that we can look from here and not see snow. In this gap is the mouth of the Sacramento river and valley to the Pacific. In this belt of country it has been as warm during the winter as April is generally in Wis. or Illis. No snow or frost in the ground except on the north side of steep hills, and there scarcely a crust. This is a very remarkable phenomena that a region of country almost surrounded with snowy mountains should be regularly warm. If any of the neighbors ask any information from this country you can say to them that they had better stay at home if they are doing anything there, if not they hazard but little in coming here. Give my love to the children and accept the same.

F. M. McDiarmid

Georgetown, California,
February the 9th 1851

My Dear wife,

I am surprised Constantia that we have had but one letter from you dated the first day of Sept. last. There may be a letter or letters for me at Coloma. I have sent there the middle of last month and not since. We have been digging about here with little or no success. We have made just enough to board us, and no more. The gulches and canyons in this part have been all dug over, leaving occasionally blocks between the holes dug. Upon these there are five or six feet of earth shovelled up from the other holes. A person may clean off one of these blocks and sink it from eight to 14 feet and then find it drifted under from the other holes. Here we see no probable prospect of making anything more than *small* wages. And as soon as I become satisfied that there are no better places than those I have seen I shall have you make a sale of stock and send me money that I may return. *But before I do this I am determined to be sure that there are no better places than these I have seen.* Last Tuesday Mr. Baldwin and myself started for Nevada City. We went to Spanish Barr upon the middle fork, and struck for Barns' Barr upon the north fork of the American river nine miles from the Spanish Barr. Here we crossed this stream in a peroge or canoe, for which we paid fifty cents a piece. We stayed all night, suppers and breakfast $5. There is a foot bridge over the first or middle, for travelling over which they charge one shilling for foot men and fifty cents a head for horses mules cattle etc, etc. The banks both up and down these rivers are two or three miles long and so steep that people cut bushes and the tops of trees and tie them together upon which they fasten their packs, and then with a rope tied to the bushes they slide them down like a slay. This slay or bushes *well loaded,* holds a man back and prevents him from dashing down the mountain to the great dangers of his brains and neck. The beds, and barrs of these rivers have been very rich. But the streams have been dammed, and the water turned in dug races along the foot of the mountain, and the beds or bottoms of the rivers completely dug over. The Barrs also are turned up side down. These rivers are very rapid and discharge an immense quantity of water although they are not as wide as Rock river at Rockford. From Barnes Barr we went to Illinois town eight miles from the river; here there are 10 or 12 houses, and some mining. Just before we came to this town we struck the road leading from Sacramento city to Nevada. Here we took dinner for $1. a piece and started for Nevada 16 miles which we reached a little after dark. Three miles from Illinois town we crossed Bear river, about the size of the Peakatoniea. Nevada is situated upon Deer Creek and is by far the largest place that I have seen in the mines. It contains a population of from 10 to 20,000. The ground about this place for a large distance has been very rich, and is turned up side down. At Grass valley six miles from Nevada they have steam and water machinery for grinding up the quartz rock; for which they give $30 per ton. I have borrowed from my old friend Vic Turner $100 and $100 from Baldwin and intend to start with Baldwin, Chitton, Col. Collins, Bennett, Million Wash Parkison and others for Clamoth and Scott rivers over the coast range of mountains in Oregon some 350 or 400 miles from here. We are to meet at Marysville the highest steamboat landing from Sacramento city on the 19th instant. We shall pack over with mules. I have bought one today for $100 and shall try to buy another tomorrow. They are very high. Do not be concerned, My Dear, about us, even if you get no other letter during the summer. I may have no chance to send any. If I do be assured that I will write you. Write me Constantia every month; direct to Sacramento city. Give my love to the children and accept the same yourself.

Yours affectionately F. McDiarmid

Weaverville California April the 1st 1851
My Dear wife,

I can only say to you that we are up in the mountains some 280 or 300 miles from Sacramento City in good health. We are with Wash Parkison, Notts and 9 others in draining a flatt. William and Rufus Henry, Martin and 9 others are engaged in a similar work just below us. Mr. Baldwin, Loveless, John Stuart,

Mathew Chilton, Sam Warren, Stephen Ensley and six others are in a similar work just above us. We do not know yet whether it will pay or not. As soon as I can determine whether I remain here for the summer or not I will write you. Send your letter to Sacramento at present. I have received but one letter from you dated the 1st of Sept. 1850. Have you sent any other, if so, to what place did you direct them? Another package will finish my tiresome journal. The next express from here to Sacramento City I will send it; which will be the first of next month. I hope with the return Express which will be about the 24th of this month to hear from you. Permit me my Dear, through you to care the children about the mill or water for fear of accidents. If I could embrace you my Dear and sport with my fond children once more I would feel a happiness never before felt. With respect I remain your affectionate hus.

F. McDiarmid

Weaverville Cal. April the 14th 1851

My Dear wife,

We are still at work upon the ditch drifting into the flatt upon both sides of the ditch. This tunneling is done with less labour than stripping or ditching, as the covering is from ten to 15 feet deep. Saturday we came across two small holes or pockets in the rock, from which we got $73 the day before we only got $7. The gold is generally coarse, some pieces weighing from $5 to eight dollars. This coarse gold leads me to think that there is some considerable quantity deposited somewhere in the flatt. The other two companies are discouraged with their appearances and are about to abandon the work. We shall be further satisfied with ours before we leave it. If we fail here, we may try some other places in this neighborhood. I would like to examine the head waters of the Sacramento river. From the appearances of this country, at a distance, I think it as favorable for gold as any in Cal. And from the warlike spirit of the Pitt indians, who are very numerous, and occupy that country, it has never been much examined. But I am tired travelling over this mountainous country. In our journey from Greenwood Valley here we have crossed a no. of the most important rivers in California. First the middle fork of the American river which we crossed at Murderers Barr. By ferry for $5. Secondly, the north fork, which we crossed by ferry at the mouth of this river or its junction with the middle fork. These are both rapid, angry, rocky, streams. These streams are closely confined by immensely high mountains. The next is Bear River. This river we crossed upon the Sacamento valley, consequently this current is less hurried and banks low. It is 15 or 18 rods wide and three feet deep, with a beautiful bright gravelly bottom. The country between this river and the mountain is steril and barren, with scattering scrubby oak upon a soil of burnt, broken stone, over which a light quantity of earth had been carefully sown. From this river to the Uba the ground is level and barren. This is a wide stream of water with a very swift current, yet it is forded. One mile from this ford upon the west side of the river is Marysville a pleasant, brisk, business town. It is the head of navigation upon the Uba river. A small class of steam boats ply regularly between this place and Sacramento City; a distance of 55 miles. Two miles below Marysville is Uba City. It is situated in the forks of Uba and Feather rivers. Marysville is the most pleasantly situated and is doing the most trade. From this place we went to Feather river; up this river we travelled for several days. There are places on flatts upon this river that appear favourable for cultivation. With the liability to deep inundation some seasons. There is a general scarcity of timber in the valley. In some places upon the bottoms of the river there is an abundance of scrubby oak. But this timber is very little of it fit for rails, nothing more than firewood. We have crossed Feather river, for which we paid one dollar ferriage for each animal and fifty cents for footmen. From this ferry we go west for

Sacramento river; up this stream we have travelled several days. The country as we go up this river becomes higher and more uneven as we approach the upper end of the valley. From where we first struck the valley to Reddings Springs, at the upper end, a distance of 275 or 300 miles the road is very good. From Reddings to Scott, Salmon, Trintity, and Clamoth rivers every thing has to be packed. The day that we came into Weaverville the people had just finished scoreing a fellows back with a rope for stealing mules. The next day they treated the back of another fellow to the end of a rope for stealing. I must now give you a history of a trial that happened on Friday last. A person was arrested by the crowd charged with *salt* and *battery* with intent to *kill*. Maj. Blythe of Dodgeville, John Stewart of Greens Prairie and myself were called upon as jurors to try the offender. I was induced to act upon the case for these reasons. For fear that he might be punished by persons who delighted in cruel punishments for the sake of seeing such exhibitions. I took part in the play. The jury was called, and ordered to be seated upon a large pine log in the midst of the infuriated crowd, in the street, by a man who assumed the powers of a court of judge. Witnesses were called and *sworn* by this man to give *true testimoney* between the state and prisoner, who had council. After many *question* of *law* were *settled* by the *judge* and *council* for the prisoner, the testimony was given in and the jury ordered to take the case. Another *log* was picked out for *us* upon *which* we were ordered to set until *we* agreed upon a *verdict.* Some were for hanging, some for whipping and branding, and some in favour of making him pay a fine, and damage to the man whom he had injured by striking him two blows upon the head with a pick. Maj. Blythe, Stuart, Lathrop, and myself could not consent to any of the punishment recommended except the last which the rest would not agree to. At last we concluded, rather than set out upon the *log* all *night* to give him over to Colonel Johnson who was lately elected justice of the peace but as yet had not been sworn in, with our verdict, that he was guilty of salt and battery. The justice took his bond, with security, for his appearance at the county court. He was then set at liberty. Thus ended one of the great trials in California. As I am writing there is a dark thunderstorm dashing around us, with the lightnings clashing and the thunder ratling!!! Strange visitors for California!!! Mr. Warren may mail this package at Sacramento, or in some of the Post offices on the way. The Express for this place will return from Sacramento City on the 24th, when I am in hopes of getting a letter from you. Tell John Crellen that Charley made a will, although verbal, and willed him one half of all his property and his brother the other half. John had better pay the taxes regularly upon the land; also the back taxes if any due. Chilton, George Cee, and another person are witnesses to his will. I will send you another letter the first of next month. My Dear remember me kindly to the little boys and to sis.

Yours affectionately
F. McDiarmid

Eel River, Trinity County, Cal.
October 6th 51

My Dear wife,

I began another letter for you at Weaver soon after I sent you my last, which I finished here, this letter or package with two others, one to my father and one to Robt M. Still each of them containing three sheet of paper setting forth every thing that has occured to us of any importance in California, a careful description of the country, its resources, the population of the cities, the character of the people, the progress of agriculture, in fact all the great elements of the state were carefully and impartially weighed by me while travelling through the state. These three packages I sent to Humbolt Post Office by a *Dutchman* who put them in his bosom, and went off the road in the grass hunting Elk and lost the letters. When I recollect that it was through the sleepy carelessness of a *dutchman* that we lost our *horses,* and by the carelessness of a Dutchman that I have lost these letters I think it necessary to trust them no more, while in California at least. I am sensible that you think it strange that you have not heard from me since June. I left the hot scorching mountains of Weaver for this place on the 3rd of June, and reached this valley on the 15th by the way of Union, Eureka, and Humbolt. Union is a small new town built at the north end of Humbolt Bay upon a beautiful piece of sloping prairie covered with an immense quantity of red clover in full blossom. This prairie twelve or 15 miles long and from six to ten miles wide is surrounded with an immense forest of Red wood timber. Varying in size from one to thirty feet in diameter and from one hundred to five hundred feet in length. I did think that I had seen in the pine and hemlock forest upon the River Tames, Trent and Credit in Canada large trees; but I must say in all seriousness that they when compared wih the Red woods of California mere *sprouts* or *whips* nothing more. Eureka is located upon east side of the Bay, mid way between Union and Humbolt which is built nearly opposite the entrance of the ocean into the Bay. It is a pleasant location and from an elevated smoothe bench of ground you can see vessels and steam boats at a distance. Eel River empties its clear water into the ocean at the south end of the Bay. The Bay is from 15 to 20 miles long and from 2 to 6 wide. The best Salmon that I have ever seen were caught from Eel River and the Bay. Some people say that during two thirds of the year this river can be navigated by steam boats for the distance of twenty miles from the ocean. At present however, it can be forded with horses safely. Horatio and I with some 30 more are making claims up this river. We have selected places 12 or 14 miles up the river from the Bay. The soil is in appearance similar to the prairie soil of Illis. and Wis. Rich and beautifully faced. Producing one class of flowers after another the year round. Some two weeks ago the indians set fire to the grass which devoured every vestige of stuff from a large scope of high dry ground. In this short time, and in a very unfavourable period of the year for vegitation to grow, this place affords as good grazing as I have ever seen. Our two mares and two mules feed upon it continually. For raising stock I believe this valley has no superior in the world. I can say of it what I cannot say of any other place I have ever known, it has an immense quantity of wild clover, both red and white; timothy and other tender grasses the year round; no Deer *flies,* no big *green* heads to suck away their blood and life, no wintry blasts and chilling snow storms to wear away their flesh. Thus we save an immense amount of money in many other countries laid out in stables, barns, sheds, and other conveniences for stock during the winter together with a large amount of labour laid out in procuring provenders in the summer and feeding out in the winter. Constantia, I do contemplate moving to this country without thinking seriously of the dangers and the difficulty of such a journey either by land or water. But great and alarming as these dangers truly are I would not shrink facing them all for the prospects of living easier here and the advantage that it will be in the future to our children. Cows are worth from $50 to one hundred dollars a head. Oxen from one hundred and fifty to three hundred dollars a yoke. If I had my family here with what stock I have, say nothing about the land, I would consider myself much better off than I am now, and my chances for making money far better than they are at present. In both your letters on January and March you say, ''I think I never experienced a more disagreeable winter.'' I must say my Dear that it is disagreeable to bear the intense heat of the summer and piercing cold of the winter in the country. If I come to California I would drive all the cattle and horses that I could buy and winter at Salt Lake; in this arrangement I would have the fall grass to Salt Lake and the spring grass over the mountains of California. I have built a small substantial frame kitchen 15 by 14 feet in size, with good floors tongued and groved. Give my thanks to the little children, and especially to Adelbert for his remembrance of me.

Eel River, Trinity county Cal. Dec 8th 51

My Dear wife,

I avail myself of a few moments while the Steam Boat is detained unloading her freight to write to you. We have our healths exceedingly well, and are busy about our claims and are now preparing to build a two story farm building at the Bay 40 feet long and 26 feet wide. We can rent the base part of the building for a store to a merchant who is trading here to good advantage. I feel very anxious to hear from you particularly so, from the fact that we are making improvements here, anticipating this country as our future home. For the reasons which I gave you in my previous letter. The longer my stay here the more I like the climate and admire the country. What little vegetable stuff that was planted upon the ground plowed badly, or broke up with the old "Cary" cast iron corn plow in June is turning out remarkably well. Potatoe patches when the land was only ½ plowed, or when one furrow was thrown up on the solid ground, turn out a bucket of potatoes from every two to three hills of as good, if not better potatoes than I have ever seen, weighing from one to four pounds each. And that without hoeing or any care after planting. I have taken from Mr. Coopers claim near ours, on the Eel River, to the Bay, two potatoes weighing 2 ½ lbs. each which I place on board of the steam Boat for Oregon as potatoe specimens or *Murfy* curiosities for the *Irish* of that land to wonder at, and laugh over. I have seen tomatoes grown here of an astonishing size. Also Beets, carrots, parsnips, cabbages, turnips, and rootabagoes of an astonishing size. Beans and peas grow and bear well. Corn, no sucessful experiment has been made. A few hills were planted by Mr. Deprue at the Bay, and one kernal by Mr. Davis upon Eel River. The stalk upon Eel River as well as those upon the Bay had as strong and thrifty a growth as any corn stalks that I have ever seen. But unfortunately after the ears became well set the mules broke in and devoured them. From this I infer that as good corn can be raised here as in Wisconsin, Illis., Ohio or Indiana. But one of the great substantial advantages lies in the mildness of the winter. At present the flowers of a delicate sort are still plenty over the prairie, and the grass is as fresh and green as it was in May and June. We have had three frosts since the beginning of this month; not hard enough to kill the nettles. The people are busy digging the potatoes. If, My Dear, you think favourable of making a move to this country and in my absence any offers of selling present, encourage them for I prefer this country Constantia at almost any sacrafice of property. If those little seedling appletrees were grafted this spring they would be a very fine size to box up and ship to this country. They would be valuable property here. The settlers here are sending to Oregon for fruit where they pay from one to five dollars a tree. If I move to California I shall drive all the young stock that I can get. If we had now here as good a lot of blooded stock as we left they would fetch in cash $100 a head; *quick,* all round. As yet I have only three letters from you since I left home. If you fall in with my views of risking a change of place let the attention and care of your father be confined soley to the raising of the stock; for they are better than cash here. If he requires any assistance let him call upon my brother. I would much sooner pay him than a stranger. There is now a post office established at Humbolt Bay. Please direct your letters to ''Humbolt Bay, Trinity County California''.

I and Horatio have rode from Eel river today and I am writing these hurried sentences when all are fast asleep about me. I know my Dear that you have written me letters that I have not received but do not cease to write; *write the oftener.* Remember me to Sis and the little boys, your father, mother, and my brother and family. With the greatest anxiety to see you my Dear, I must bid another letter good bye.

F. McDiarmid

COLOPHON

The Finley Mc Diarmid, LETTERS TO MY WIFE, was printed in the workshop of Glen Adams, which is located in the sleepy country village of Fairfield, southern Spokane County, Washington State and one township removed from the Idaho line. The text was keyboarded by Teresa Ruggles using a Compugraphic Editwriter computer photosetter. The typeface used is 14 on 16 Garamond with the running heads and page numbers in Baskerville Bold. The camera/darkroom work was by Heather Paulson using a Companica 660C DS computer driven camera. The film was stripped by Susan Cardwell/Paulson who also designed the cover/title and colophon pages and made the printing plates. The sheets were printed by Trevor Del Medico using a 28 inch Model KORS Heidelberg Press. The sheets are 70 pound Quasar offset. Folding was by Garry Adams using a 26x40 inch Baum Dial-O-Matic folding machine. Garry Adams also assembled the folded sheets. Hardcase binding is by Al Chidister of Oakesdale, Washington. Paper binding is by Glen Adams and Garry Adams using a Sulby Mark II adhesive binding machine. This was a fun project. We had no special difficulty with the work. This is the 609th title to come from Ye Galleon Press.